Agricultural Administration Unit

Occasional Paper 11

Managing Agricultural Research for Fragile Environments: Amazon and Himalayan Case Studies

John Farrington

and

Sudarshan B Mathema

Overseas Development Institute

British Library Cataloguing in Publication Data:

Farrington, John
Managing Agricultural Research for Fragile Environments:
Amazon and Himalayan Case Studies
1. Agriculture. Research projects
I. Title II. Mathema Sudarshan B III.

ISBN 0-85003-139-7

Published by the Overseas Development Institute, Regent's College, Inner Circle, Regent's Park, London NW1 4NS

Printed and typeset by the Russell Press., Nottingham

Contents

List of Tables, Figures and Maps

Annexes

Figures

Maps

Preface and Acknowledgements

Substantial increases in crop and livestock productivity have been gained in agro-climatically favourable areas of developing countries through approaches to agricultural research which differ little from those long practised in the North. Large areas characterised by low and unreliable rainfall, poor soils and hilly topography have been largely by-passed in this process. Yet, the majority of rural population in many countries seek livelihoods in these areas, and evidence is emerging that innovative approaches to research can generate substantial and sustainable increases in productivity there.

Drawing on case studies form two widely-differing and highly fragile environments — the mid-hills of Nepal and the Amazonian lowlands of Bolivia — this book explores how agricultural change sensitive to environmental conditions has been introduced, seeking to draw wider lessons from the methods and institutional forms that were introduced.

The authors are indebted to the Directors and staff of the British Tropical Agriculture Mission (Bolivia) and Lumle and Pakhribas Agriculture Centres, (Nepal) for their support in the preparation of case study material and for comments on an earlier draft. John Howell, Mary Tiffen, David Gibbon, Ramesh Khadka, Penny Davies and others at BTAM, PAC and LAC kindly also commented on drafts. In Nepal, K R Regmi and B P Upadhaya provided research assistance. Kate Cumberland tolerantly and skilfully typed successive drafts.

The authors are indebted to the Ford Foundation (New Delhi) which provided support for fieldwork in Nepal, and to the British Overseas Development Administration. What follows should not be taken to reflect the views of either of these organisations.

Glossary

ARPP	Agricultural Research and Production Project (Nepal)
ASC	Agricultural Service Centre (Nepal)
BTAM	British Tropical Agriculture Mission (Bolivia)
CIAT	Centro de Investigación Agricola Tropical (Bolivia)
CIAT	Centro International de Agricultura Tropical (Colombia)
CIP	Centro Internacional de la Papa
CDR	Complex, diverse and risk-prone agriculture
CSP	Cropping Systems Programme (Nepal)
DoA	Department of Agriculture (Nepal)
DC	Development Corporation (Bolivia)
FFT	Farmer's field trial
FRSDD	Farming Systems Research and Development Division (Nepal)
HMG(N)	His Majesty's Government of Nepal
IARCs	International Agricultural Research Centers
IBTA	Instituto Bolivianode Technología Agricola
ICIMOD	International Centre for Integrated Mountain Development
ICP	Integrated Cereals Project (Nepal)
IRRI	International Rice Research Institute
ISNAR	International Service for National Agricultural Research
JSSCEN	Joint Services Scheme for Civilian Employment in Nepal
KHARDEP	Kosi Hills Rural Development Project (Nepal)
LAC	Lumle Agriculture Centre (Nepal)
ldc	developing country
NARCC	National Agricultural Research Coordinating Committee (Nepal)
NARS	National Agricultural Research Service

NARSC	National Agricultural Research and Services Centre (Nepal)
NGOs	Non-governmental organisations
ODA	Overseas Development Administration (UK)
PAC	Pakhribas Agriculture Centre (Nepal)
PPVT	Pre-production variety trial
RCC	Research Coordinating Committee (Nepal)
SERED	Socio-Economics Research and Extension Division (Nepal)
STIP	Seed Technology Improvement Project (Nepal)
USAID	United States Agency for International Development

Abstract

This book arises from three sets of concerns: that much agricultural development has been insensitive to environmental issues in fragile areas; that conventional approaches seeking more sensitive and sustainable technologies for small farmers in these areas have generally been unsuccessful, and that, although recent participatory approaches to technology development have been successful on a limited scale, substantial problems remain to be overcome if they are to be institutionalised more widely. The book draws on detailed case studies of successful and institutionally innovative approaches to research, dissemination and feedback in two widely-contrasting environments — the hills of Nepal and the lowlands of Bolivia. The case studies are set against a broader overview of agricultural research structure in the two countries. Major conclusions are: that many of the principles of participatory approaches can cost-effectively be institutionalised by the public sector to produce adoptable and sustainable small farm technologies; that agencies outside government are an important source of research and extension and that in these areas and government should find ways of working with them; that, in order to be effective, donor involvement should concentrate less on short-term inputs and more on the long-term involvement of technical cooperation staff; that scarce government resources can be utilised more effectively if given thematic and geographical focus, and the pressures towards setting up many research units of sub-economic size resisted;that no blueprint to successful institutional design or technology development exists — local solutions must be developed to fit local contexts; that a high degree of decentralisation of day-to-day management is needed if such solutions are to be found; and that the role of the international agriculture centres and of national centres in finding solutions for fragile areas has been limited, so that collaboration between institutions focusing on fragile areas and these, although highly desirable in principle, should in practice be closely limited to areas of likely mutual benefit.

Map 1: Santa Cruz Department: zones of production

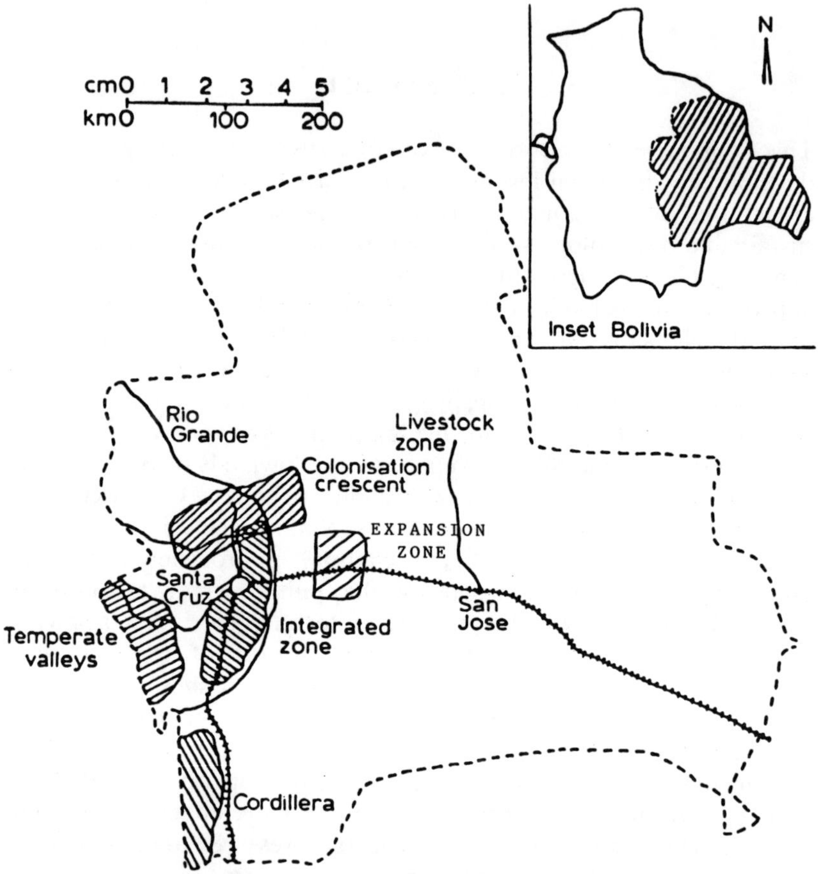

Map 2: Nepal: Agro-ecological regions, and command areas of LAC and PAC

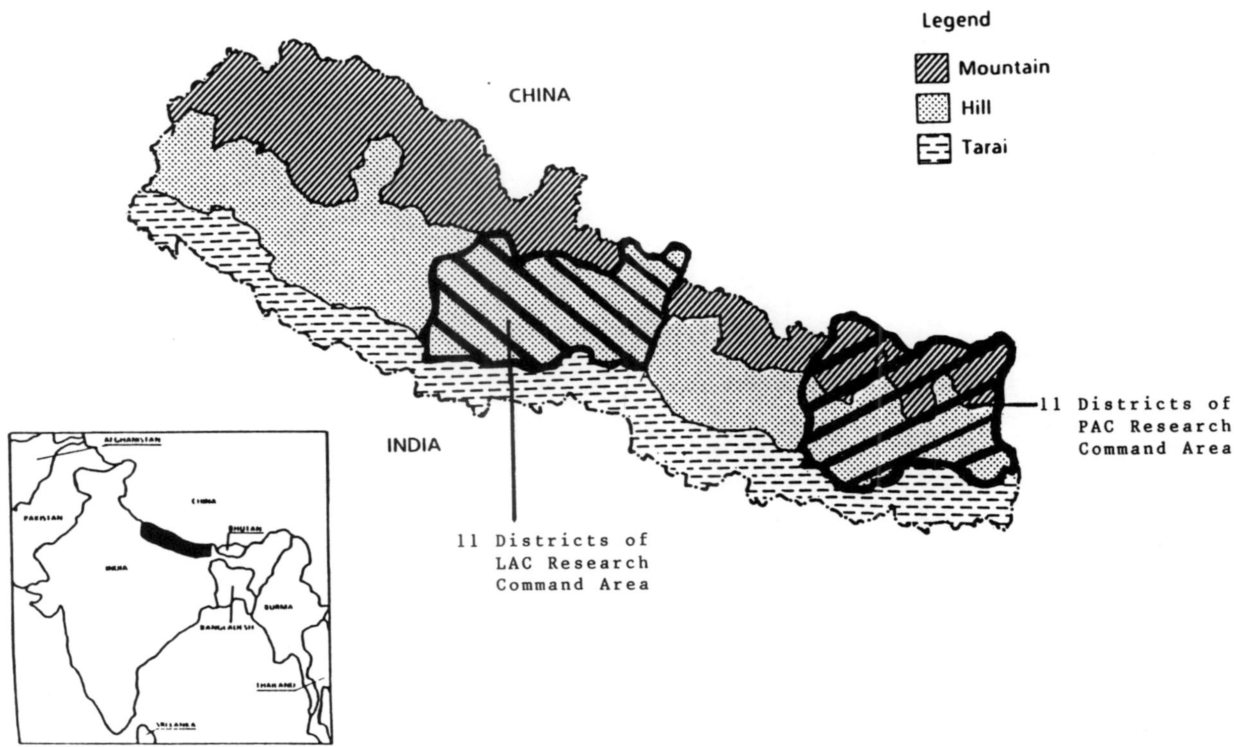

1
Introduction

Conceptual framework

The central question addressed in this book is how best use can be made of resources available for agricultural research and extension in those areas characterised by some combination of low and unreliable rainfall, poor soils and hilly topography. At least 500 million people[1] — ie. over 25% of the rural population in less-developed countries (ldcs) — depend largely on agriculture (including annual crops, trees and animals) for their livelihoods there. Such areas have been characterised as 'complex, diverse and risk prone' (CDR) (Chambers et al. (eds), 1989). They lie outside other areas — usually reliably rainfed or irrigated — which have been the focus of agricultural development efforts, including the 'Green Revolution', over the last 3 decades.

Discussion of *how* research should best be organised for CDR areas must first respond to the argument that, because of the low inherent productivity of these areas, research resources are best allocated elsewhere — eg. to smallholders in the better endowed areas. Whilst in theory it is not impossible to arrive at an economy-wide investment model that will indicate the optimal pattern of allocation of research resources, even to such sub-sectors as the CDR areas, in practice the development of such a general model is virtually impossible depending, at a minimum, on:

- knowledge of the marginal returns that will be generated by investment of additional resources, including those for research, in each sector and sub-sector
- agreed assumptions about society's preferences for income now as against income in future time periods

1. Estimates vary, partly because of the problems of defining these areas precisely. Mellor (1988) suggests that over 300m people live in poverty in areas of 'low agricultural potential'. Others (including Chambers, pers. comm.) have argued that the number of persons dependent on agriculture in these areas is close to 1bn.

- agreed assumptions about the desired patterns (spatial, interpersonal) of distribution of future income.

This minimum list of requirements poses severe difficulty for macroeconomic planning even in industrialised countries. For ldcs, the difficulties of arriving at agreed assumptions and quantifying responses to investment at sectoral or sub-sectoral level are so severe as to be virtually insuperable. Measuring the returns to agricultural research has, to take one example, raised numerous methodological difficulties even in the North where reliable data are available (Wise, 1981); these are all the more severe in ldcs.

Given these difficulties in determining the volume of public[2] research resources that should be allocated to CDR areas within the context of a general allocative model, the alternative adopted here is to review more circumstantial evidence on whether current investment levels are adequate. Three broad sets of factors are relevant:

First, economic conditions in ldcs are now very different from those faced by presently industrialised countries in the last century: ldcs have a higher proportion of the labour force in agriculture; labour-saving devices in industry have reduced its employment-generating capacity, and overall population growth rates are much higher than in the last century (Lewis, 1979). Attempts have been made to interpret these conditions in terms of a single indicator: the number of years likely to elapse before a turning point at which the absolute size of labour force in the agricultural sector begins to decline. Despite methodological difficulties, analysis of this kind provides broad guidelines: estimates presented for selected countries in Table 1 for standard and rapid rates of overall population growth and for differential rates of labour absorption by the industrial sector indicate that this turning point has already been reached for some countries (Brazil; South Korea). However, even under the most favourable assumptions, it will not be reached in India for at least 15 years, in Bangladesh for 20 years and in Kenya for 50 years. Given the fixed land frontier in most countries, new technology will be needed to enhance the productivity of existing agricultural land in most areas if the living standards of this increasing rural population

---

2. Unless stated to the contrary, the discussion throughout excludes research and dissemination conducted by the private commercial sector. The impact of this is small and restricted almost exclusively to well-endowed areas (Pray, 1987).

are not to decline. In some countries, and particularly in CDR regions, the issue is likely to be more one of reversing the decline in productivity caused by depletion of the resource base.

Second, there are problems of quantifying the research resources allocated to CDR vis à vis other areas, and the respective returns that they have generated. Published statistics to permit cross-country comparisons of resource allocation and productivity between CDR and other areas do not exist. The problems of delineating CDR areas, of measuring resource allocations to them and the problems introduced by measuring productivity in conventional terms (such as per unit area) whereas in CDR areas, a high proportion of useful biomass is generated from common property resources, combine to make the compilation of such a statistical series particularly difficult.

There are also related difficulties: it may, for instance, be possible to estimate for specific areas in a given country that the rate of growth in value of output attributable to research over a given period — say, the last decade — has been twice as high in well-endowed as in the CDR areas. According to economic theory, this only tells us ex post that the incremental resources made available for research over that period should have been allocated in the same proportions, and this assumes away the important problems of detecting what proportion of productivity increases is attributable to what research, and of the time lags between research, adoption and productivity growth (Wise, 1981). The data series tells us nothing about where the highest potential returns to research expenditure will be for the future, nor is it directly translatable into prescriptions concerning the overall (as opposed to marginal) level of funding.

Third, difficulties arise in comparisons of research productivity between CDR and other areas. In the latter, the contribution of new technology to enhanced productivity can be measured in terms of such conventional criteria as the returns to land, labour or capital. In CDR areas, the natural resource base may be in long-term decline eg. as a result of population pressure, so that such conventional measures will give an underestimate of productivity if they postulate a steady-state baseline. In other words, a fundamental objective of research is to sustain or enhance the productive base, so that any success in doing so should be included in the returns to research. Further, much of the useful biomass produced in CDR areas comes from off-farm resources (forest, grazing) so that measures of the returns to research should also include these. Finally a fundamental objective of the more participatory modes of research in CDR areas

Table 1: Projected growth of agricultural labour force in selected countries under various assumptions

Country	*Non-agricultural employment growing at 4% a year*		*Non-agricultural employment growing at 2% a year*	
Assumed overall Population growth:	Standard	Rapid	Standard	Rapid
Kenya (78% of labour force in agriculture, 1980)				
Turning point[a]	2045	2040	2065	2045
Thailand (67%, 1980)				
Turning point	2010	2005	2025	2025
Bangladesh (74%, 1980)				
Turning point	2030	2010	2060	2040
India (69%, 1980)				
Turning point	2005	2005	2030	2025
Malaysia (50%, 1980)				
Turning point	2000	2000	2015	2015
Egypt (50%, 1980)	2000	2000	2025	2020
Brazil (47%, 1980)				
Turning point	1970	n.a.	n.a.	n.a.
Korea (66%, 1960)				
Turning point	1960	n.a.	n.a.	n.a.

a The year when the agricultural labour force starts to decline.
n.a.: Not applicable
Source: Birdsall (1985, Table 6)

is to strengthen farmers' own capacity for experimentation; evaluation and dissemination. To the extent that they succeed, the additional capacity represents a return to research investment.

In the absence of allocative prescriptions based on coherent conceptual models and sound empirical analysis, the best that can be done is to examine the validity of common assertions concerning the allocation of research resources.

A set of assertions favouring the continued allocation of a high proportion of research to well-endowed areas is made by Mellor (1986), though he subsequently (1988) retreated somewhat from this position. Briefly his argument was that high rates of productivity growth in well-endowed areas are attributable, with little lag, to identifiable research; that the same growth rates and causality relations will hold for the future; that further high levels of research in well-endowed areas are therefore economically justified and will create technologies that absorb labour from the stagnating CDR areas, and that research for the latter should be a much lower priority, taking place in an unspecified 'second stage', financed by taxation of the surplus generated in the better endowed areas.

Whilst there is some evidence that growth through new technology in well-endowed areas leads to increased employment in agriculture-related activities, some (eg. Chambers et al (eds), 1989) have argued that the same causality between research, productivity and employment will not hold for well-endowed areas in the future, where problems of salinity, resistance to pesticides and environmental damage through nitrate and pesticide pollution are beginning to emerge, thus increasing the attractiveness of CDR areas for investment of research resources. Other arguments have been advanced in favour of increased allocations to CDR areas: the returns to publicly-funded research in these areas can be much higher for the future if it adopts organisational forms and methodologies capable of understanding and strengthening the systems interactions, and therefore, overall productivity of CDR agriculture. There is also evidence that it will be higher if it seeks collaboration in field trials and dissemination with a range of local organisations (Thiele et al., 1988 — see also below).

There is also evidence to demonstrate that funding in some CDR areas has been too low and fragmented to allow a critical minimum mass of effort to be established (K.C. et al., 1987; Sharma et al., 1988). Undoubtedly, though, one of the most powerful arguments for increased allocations to CDR areas is that in many countries population levels there will tend to increase over the coming decades (Table 1), and the productivity decline of the resource-base observed in many areas must be reversed if livelihoods are not to be ended. Policy to arrest the widening gap in incomes between CDR and other areas has, in some cases, addressed these concerns by specifying a shift in agricultural research resources towards CDR areas (Government of India, 1985).

Overall, the view underlying this book is that, in most countries, well-endowed agriculture is likely to be more important in absolute economic terms and exhibit faster rates of growth than CDR agriculture, and should therefore receive the bulk of public funding for agricultural research. There are, however, strong arguments for a substantial minority of funds to be allocated to CDR areas, sufficient to give a higher absolute level of funding than hitherto.

Two further questions arising from this opening discussion are introduced here, since they form the core of the analysis and case study material presented below. These are, first, whether optimal use has been made by publicly-funded services of the sparse resources that have been provided to CDR areas; the second is whether all the funding for agricultural research and extension in CDR areas must necessarily come from government.

In response to the first question, a wide range of evidence has been accumulated to indicate that research structures, organisation and methods in the public sector have been inappropriate to the needs of CDR areas (Chambers and Ghildyal, 1985; Chambers and Jiggins, 1986). Structural difficulties include the dependence on centralised and, therefore, distant decision-taking, not only for research proposals but on matters relating to the day-to-day requirements (supplies; maintenance; repairs) of operating research facilities effectively in remote conditions. Some of the organisational and management disincentives for working in CDR areas include a requirement to produce publishable work, and poor facilities and low levels of allowances for working in difficult postings. Numerous problems in methodology have also been noted: research agenda have been defined by scientists whose perceptions of complex and diverse local conditions are limited; technology has been tested and evaluated under atypical practices and conditions, and the predisposition among many scientists — often reinforced by training in western institutions — towards straight-row, clean-weeded pure stands of arable crops limits the scientist's capacity to relate to the CDR conditions under which mixed stands, relay cropping and combinations of crops and trees are known to be highly productive (Chambers and Ghildyal, 1985). Successful introduction and adaptation of technology in these areas has typically been carried out by farmers themselves, NGOs, universities and, occasionally, decentralised publicly-funded research institutes. It is characterised by researchers' acceptance of farmers as equal partners in research, a willingness to learn from their practices, understand the opportunities and constraints they face, and bring these to bear on the research agenda. This participatory mode characterises successful research through the entire process of problem identification, experimentation, evaluation and dissemination.

As far as the second question relating to the source of funds for research in CDR areas is concerned, evidence is emerging that actors other than government have generated adoptable technologies. First, farmers themselves have demonstrated extensive capacity for experimentation (Sharland, 1989; Richards, 1986), although not all farmers display identical inclination and capacity to innovate (Biggs, 1980). This evidence is consistent with the extensive farmer experimentation noted in Europe and the USA before research was institutionalised and is contrary to a common view that CDR farmers are so conservative and risk-averse that any new technology has little prospect of adoption. Farmers' capacity for experimentation represents a substantial resource which, if properly harnessed, can complement the resources provided by government. Second, within a 'tripartite' conceptual framework (Röling, 1988), it has been argued that there are advantages in incorporating local organisations such as NGOs, farmers' organisations and cooperatives as an intermediary between researcher and farmer.

The extent of involvement by NGOs and farmers' organisations in agricultural technology development has tended to be underestimated, not least because those involved are practitioners with limited incentive to document their experience. Nevertheless, case studies are now being assembled and written up under an ODI research project[3] and from a preliminary review (Farrington and Biggs, forthcoming) it is clear that a large number of NGOs and farmers' organisations have successfully introduced and disseminated agricultural technology, particularly in CDR ares, though poor communication among these organisations, weak institutional memory and limited professional capacity have in some cases led to 'reinventing the wheel', missed opportunities and undue attention to non-viable technologies. An option for the larger NGOs in addressing these shortcomings is to appoint technical advisers on a country or Regional basis (as CARE International has recently done — Sumberg, pers. comm.). Links with professionals in public sector agencies (whether national or international) are an option being taken up increasingly by NGOs and farmers' organisations. Examples include the collaboration between World Neighbors and the Guatemalan research service (Bunch, pers. comm.), between universities and a number of NGOs in Chile (CIAL, 1988), between research services in Bangladesh and several NGOs (Mennonite Central Committee, 1989) and, in an example discussed in more detail in Chapter 4, between NGOs and research services in eastern

3. 'Links between NGOs and public sector research and extension agencies: unexploited potential?' — led by J Farrington.

Bolivia, (Thiele et al., 1988). Donor-assisted projects are now being designed around collaboration between NGOs and the research services (as in Sierra Leone — Sumberg, pers. comm.) and there is evidence that NGO initiatives have shifted the direction of public sector research, as with the Oxfam-US initiative on vertisols research at the International Livestock Centre for Africa (Okali, pers. comm.) and the CARE-Mazingira initiatives with the International Council Research in Agroforestry and the Kenyan Forestry Research Institute (Scherr, pers. comm.).

Overall, the resources provided by NGOs and producers' organisations and the relevance of their participatory approaches to technology design appear to have been largely understated. In some cases they are greater in manpower terms than those provided by government services for technology developed in CDR areas (Thiele et al., 1988).

As identified above, low funding is, however, only part of the problem: of at least equal importance are the challenges posed for research by the complex, diverse and risk-prone characteristics of these areas, and the cluster of problems caused by remoteness: difficulties of communication between research and extension workers and farmers; the degree of decentralisation required, but rarely obtained, in administrative and managerial responsibility; and the basic problems for research managers of ensuring that employment conditions, supplies and repair facilities are adequate to keep remotely-located research stations operating effectively. Despite the increasing literature on research methods designed to enhance the participation of farmers in technology development, and so bring the research agenda closer to their needs, relatively little work has been done to identify methods cost-effective enough to be implemented on a large scale, nor has much consideration been given to the institutional forms appropriate to implementation of these methods (Farrington and Martin, 1988).

At the same time however, substantial but largely unrecorded experience has been gained in the wider aspects of organisation of agricultural research for CDR areas. Such efforts have been driven by a perceived need to understand farmers' requirements and incorporate them into the research agenda. They have thus been participatory in a pragmatic and cost effective, rather than a purist, sense. Whilst individual aspects of these approaches, or of the technologies they produced, have been published in relevant journals, there has so far been no overall review of the approaches or attempt to draw wider lessons from them.

The purpose of this book is not to attempt a comprehensive review of such experience but, as a modest start, to present case studies of innovative approaches to research and extension for difficult environments from two countries, Nepal and Bolivia, placing them in the wider context of research organisation for the countries as a whole, and asking whether, despite the diversity of physical, administrative, economic and political conditions between the two countries, broader lessons can be drawn.

It is hoped that the lessons arising from these case studies will prompt publication of further reports of institutional innovation in agricultural research and extension. The identification and documentation of examples from Africa is a particular priority. Conceptually, case studies of institutional innovation fall broadly within what Biggs (1989) has termed a 'multiple source of innovation' model. This model challenges conventional thinking on the sources of innovation embodied in what has been termed the 'central source' model (Biggs, ibid). Briefly, the central source model holds that international agriculture research centres transfer technology to ldc national services which is then adapted to local conditions and disseminated through extension services. A reverse set of linkages is supposed to provide feedback on the performance and adoption level of these technologies so that research agendas at the centre can be modified in line with farmer requirements.

The multiple source model, by contrast, postulates that innovations in technology, research methods and institutions arise from diverse sources, and that advances in these several areas influence the speed and efficiency with which technologies acceptable to the farmer are generated and spread. Research and dissemination should therefore be viewed in the context of the historical, political, economic, agroclimatic and institutional setting in which technological change takes place.

The two case studies analysed here are, essentially, practical illustrations of the multiple source model. They have drawn on the 'central system' for what it can best provide, but have drawn more strongly on a combination of local materials, methods and institutional forms, supplementing these with resources from elsewhere within the two countries, and from donor agencies. In brief, they are characterised by strongly innovative and individualistic development of research institutions, management and methods to suit the conditions for which they cater. Broader purposes of the study are:

- to draw out general conditions for success in the organisation of agricultural research for difficult environments — and the selection of two such widely differing environments and case studies as Bolivia and Nepal is a strong test of whether such generalisations can be made,
- to examine whether linkages of the type inherent in the central source model (ie. to international centres) are important to the development of these approaches,
- to assess how far these approaches are likely to be sustainable and replicable.

The remainder of this book is organised as follows: Chapter 2 provides the institutional context by reviewing the structure, performance and limitations of national research services in the two countries. Chapter 3 then focuses on the institutional innovations that have been made to overcome the particularly acute shortcomings evident in complex, diverse and risk-prone areas. Chapter 4 draws out similarities and differences among these innovations between the two countries, and the final chapter draws general conclusions.

2
Evolution, Structure and Performance of National Research Systems

Nepal

Administration

The evolution of the national agricultural research system has been much more complex in Nepal than in Bolivia. Up to the late 1960s, the emphasis was largely on the acquisition of technology from external sources. Research therefore played a minor role relative to extension and demonstration. Its role subsequently expanded as the philosophy shifted towards development of a national adaptive research capacity.

To a substantial degree, institutional developments reflected this evolution. Following the establishment of the Department of Agriculture in 1924, a demonstration farm and a fruit nursery were opened, both near Kathmandu. Several further agricultural stations and farms were established in the 1950s and 1960s in other parts of the country. Farms were each allocated a single activity (eg. horticulture, livestock) whereas stations were multi-purpose. The functions of both were geared to the prevailing extension and demonstration philosophy: seed multiplication programmes, demonstration trials and multi-locational testing were staple activities. In parallel, village-level workers were employed under block development programmes, and USAID assisted in establishing a village development training centre in Kathmandu in 1952. These programmes were oriented to specific production targets and emphasised the extension of new technologies to farmers, and farmer-training.

Research, extension and training during this period were coordinated by the Department of Agriculture which in 1957 comprised specialist sections in agronomy, horticulture, livestock and dairy, veterinary medicine, agricultural engineering, plant pathology, entomology and fisheries. A further five new sections were opened during the first 5-year plan period (1956-1960): soils, extension, botany, marketing and farm management, and a training school. Throughout this period, exotic material for crop, horticultural and animal production was brought in, mainly by donors and foreign advisers and tested by the disciplinary divisions. Crops released included Lerma-52 and the NP series wheat varieties, Cubano maize and CH45 rice. (See Annex 1).

The agricultural demonstration and extension philosophy was strengthened in 1965-66 when USAID funded an intensive, coordinated agricultural development programme in nine of Nepal's highest potential Districts (six in the Tarai and three in the Kathmandu Valley). A further seven Tarai Districts were added in the following year. Local capacity to respond to the research and extension requirements of this programme was, however, impaired by the first in a series of major reorganisations which were to take place over the following 15 years. The Department of Agriculture was replaced in 1966 by five new Departments: Agricultural Education and Research; Agricultural Extension; Horticulture; Livestock and Veterinary Science, and Fisheries. The former DoA Agricultural Economics Section was renamed Economic Analysis and Planning Division and brought directly under the Ministry. This extensive subdivision of responsibility made it impossible for station managers to continue coordinating across disciplines as they had sought to hitherto. Yadav (1987), for instance, argues that fragmentation and duplication of effort increased as each new Department sought to bring a portion of each station under its control.

These five Departments were then subsequently merged back into a single Department of Agriculture in 1972-73. Coordinated crop development programmes characterised by multidisciplinary commodity focus or problem focus were established for paddy, wheat, maize, citrus and potatoes. A two-tier approach was adopted: the sections and stations were administered from Kathmandu, whereas commodity and discipline-specific research farms fell under one of four new Regional Agricultural Directorates. Major strategic bodies such as the National Agricultural Development Committee and the Agricultural Research Coordination Committee, established at the same time, remained largely dormant.

Further changes included (in 1979) the splitting of a Department of Livestock Development and Animal Health from the DoA, and the creation (in 1982) of three Deputy Directorates General within the DoA to take charge of research and extension in respectively: crop production; horticulture and fisheries; and extension and such technical services as soil science, chemistry, pathology, entomology and engineering. Again, experienced observers noted the increased difficulties in coordinating work effectively under these complex and rapidly-changing institutional arrangements (Yadav, 1987). Pressures for stronger adaptive research were emerging in the 1960s and 1970s as both donor programmes and Nepalese scientists returning from overseas training recognised the limitations of the demonstration/diffusion approach (Kayastha et al., 1989). Response to these pressures emerged principally at four Terai research stations which received the bulk of donor funding.

Indigenous breeding programmes were established for wheat, maize and rice, and testing procedures established for exotic material supplied by such IARCs as CIMMYT, IRRI and CIP. National commodity programmes were established in 1972 for wheat, maize and rice, each being attached to one of the main Tarai research stations. Multidisciplinary teams comprising agronomist, plant breeder, plant protection officer and soil scientist were formed around each of these programmes and linkages maintained with the respective discipline headquarters in Kathmandu. Farmers' field trials (FFT) were introduced a year later as a means of multi-locational testing of new varieties, initially in the Tarai, to which most of the material then being introduced was best suited.

New donor funding associated with substantial advances in research methodology followed from a 1975 Rockefeller Foundation evaluation which linked the lack of success of many agricultural projects with researchers' unawareness of farmers' practices and conditions. It drew particular attention to the poor prospects of successful adoption by farmers in one agro-ecological zone of technologies developed in others. One of the results of the evaluation was a request by HMG(N) to Lumle and Pakhribas Agriculture Centres (LAC and PAC) to expand their activities to include agricultural research. Another was the inception of the Integrated Cereals Project (ICP) in 1977. Apart from strengthening research within the commodity programmes, the ICP helped to set up and operate a Cropping Systems Programme (CSP) within the Agronomy Division. This had the responsibility for designing methodologies for the evaluation of crops and input technologies from commodity programmes and disciplinary divisions against a range of agro-ecological and cropping systems characteristics. The pre-production verification trial (PPVT) guidelines produced by the CSP in the early 1980s have been widely used by other development agencies conducting agriculture-related research, and influenced other DoA programmes to adopt locality-specific field trials.

A major reorganisation of the NARS was undertaken in 1985, with the aim of improving the capacity to prioritise and coordinate research activities. Following the reorganisation, the MoA comprised five major Departments, each having a responsibility for research: Agriculture, Livestock Development and Animal Health, Food and Agriculture Marketing Services, Central Food Research Laboratory and the National Agriculture Research and Services Centre (NARSC). A Research Coordination Committee (RCC) was established to guide all agricultural research.

The RCC and NARSC are newly established Departments and receive technical and financial support from the Agricultural Research and Production Project (ARPP) funded by USAID in 1985 for an initial 5-year period. Apart from its contribution to the coordination and prioritisation of research, ARPP's broad objectives are to carry forward work initiated under the ICP in strengthening research capacity in commodities and disciplines and across systems.

The RCC was established to coordinate and prioritise research activities across departments within the MoA, and between the MoA and other agencies. The NARSC was established in 1985 within the DoA to plan and monitor agricultural research programmes according to RCC policies. Two years later, NARSC was granted departmental status under the MoA and responsibility for several research programmes was transferred out of other Departments and into NARSC, including: cereal crops, pulses, oilseeds, horticulture, livestock and agricultural marketing research. NARSC's objective is to coordinate research through the sequence of planning, execution, analysis and reporting. It is responsible for the publication of results in a form appropriate to users ranging from scientists to extension workers and farmers. Technical committees under NARSC review research proposals presented on a standard protocol to ensure conformity with RCC priorities. Formats for data analysis and reporting have also been standardised, as have purchases of equipment for ease of maintenance. NARSC has also brought together previously scattered documents into a single library, and

Figure 1: Organizational chart of the National Agricultural Research and Services Center, Nepal

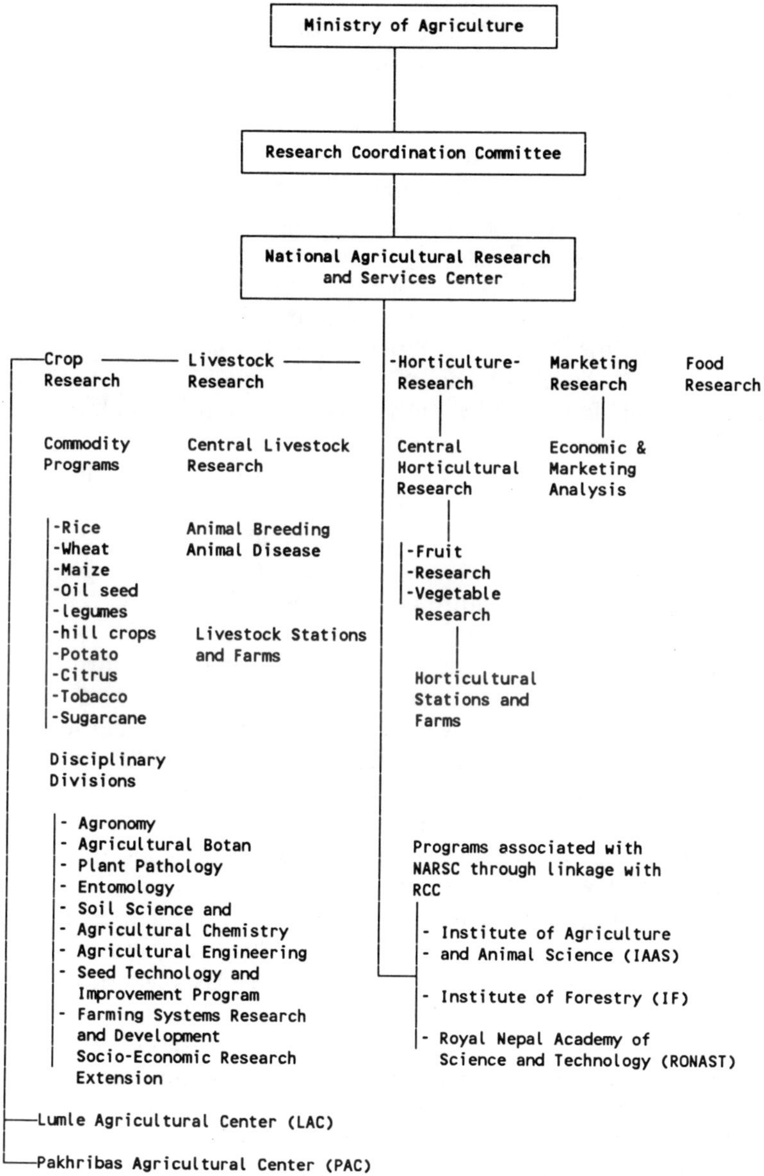

provides a central contact point for the IARCs and for donors. NARSC currently has under its management: 18 disciplinary divisions; 10 national commodity programmes and 31 agricultural stations and farms. It structure is outlined in Figure 1. The current interrelationship between NARSC, the MoA, and other agencies having responsibility for research and training in agriculture is given in Figure 2.

Finance

HMG(N) expenditure on agricultural research in the early 1970s was RS 15m/year (approximately US $1.5m) rising to Rs 19m (Us $1.7m) by the end of the decade. More recent data are unavailable. This amounted to some 19% of the agriculture sector budget, equivalent to 0.22% of agriculture GDP over the period, which is well below the 0.56% average for 51 ldcs surveyed by Oram and Bindlish (1981). Recent substantial expansion of funding especially from donor sources has raised research expenditure to approximately $4m, thus bringing Nepal close this average.

Of overall expenditure in 1980-81 (Rs 17.5m) 37% was allocated to crops research, 23% to livestock, 21% to horticulture and 8% to fisheries. The real growth of research expenditure for 1970-71 to 1980-81 has been under 4% per annum, with crops research at approximately 4%, horticulture at almost 9%, fisheries at 2% and livestock practically 0%.

HMG(N) expenditure data have been analysed for their distribution among major ecological regions (Tarai and Hill/Mountain) by Yadav (1987), who estimated that over the decade up to 1980-81 over 50% of allocations were made to the Hill/Mountain zone, concluding that '...during the 1970s, hill agricultural research was accorded a high priority' (ibid., p. 40). However, some of the assumptions used by Yadav are questionable. For instance, he includes the Kathmandu Valley in Hill/Mountain areas, whereas their conditions are more similar to those of the Tarai. Reanalysis of his data leads to the conclusion that only 38% was allocated to Hill/Mountain areas and thereby invalidating Yadav's assertion that they were a 'high priority' in HMG(N) funding.

In summary, the Hill/Mountain areas have consistently received under 40% of the national research budget but contain over 60% of rural population. It is difficult to reach a balanced view of whether this does, in fact, constitute 'under-funding' of research for the Hill/Mountain areas and 'over-funding' for the plains. The marginal productivity of additional research expenditure is likely to be higher in the more favourable and homogeneous soil, water and climatic conditions of the plains. This argues that higher allocations of resources — including those for agricultural research — have to be made to Hill/Mountain areas if spatial disparities in incomes within the country are not to widen. We argue in subsequent chapters that modifications to the organisation, methods and focus of research for these areas will produce much higher returns than those obtained hitherto. We argue that there is already a case for a higher allocation of research resources to Hill/Mountain areas. This is strengthened with the introduction of these more productive approaches to research.

Achievements

It is impossible to present here more than a broad outline of achievements and prospects. For more detailed reviews, the reader is referred to such publications as Yadav (1987); Kayastha et al. (1989) and Sharma and Anderson (1988).

Considerable success has been achieved since the 1950s in several areas:

- new genetic material in both plants and animals has been introduced, tested and adopted, at first directly and subsequently as parent material in national breeding programmes. The wide adoption of CH-45, Masuli, Sabitri, Janaki and other rice varieties, of Khumal, Amarillo and Rampur maize varieties and of Sonalika and other wheat varieties provide examples in annual crops (see also Annex 1) as does the incorporation of Holstein-Friesian blood into the dairy herd.
- research procedures have been strengthened to relate research more closely to farmers' conditions. Farmers' field trials (FFT) were introduced in 1973 and rapidly adopted across the commodity programmes. Preproduction verification trials (PPVT) were introduced in the late 1980s under the US-funded Cropping Systems Programme of the Integrated Cereals Project. Under the impetus of the Project, both PPVTs and guidelines for the establishment, conduct and reporting of trials spread rapidly. Under the CSP, 'minikits' were launched in 1977 as an aid to both research and extension. Several thousands of these kits, containing improved seed from both released and prerelease varieties and, where appropriate, fertiliser and pesticides, were distributed to farmers. Whilst these kits have strengthened the status and

Figure 2: Primary functions and linkages of organisations within the National Agricultural Research System, Nepal

Organisation				
RCC	a.	Develops agricultural research strategy. Coordinate research activities with other agencies and Ministries.		Joint-Secretary of Research, DOA acts as Member-Secretary.
	b.	Establishes research priorities for all research programs.		
NARSC	a.	Plans, monitors and evaluates the research programs conducted by the crop commodity programs, disciplinary divisions, agricultural farms and stations.		Advised by the RCC under the MOA. RCC comprised of representatives from various disciplines including farmers and research staff.
	b.	Coordinates research program in major and minor cereal crops pulses, oil seeds, livestock and pasture, horticultural crops and agricultural marketing.		Technical committees review research pro-proposals against research priorities developed by the RCC.
LAC	a.	Implements extension and training programs in specific target area in the Western Hill region supported by an on-station and on-farm research program.	a.	Supported by the British ODA
	b.	Acts as a center for basic research, foundation seed production, fruit and forest tree multiplication, and production of exotic livestock.	b.	Participates in the National Crop and FSR Workshops
	c.	Serves and supports agricultural development in the Western Hill Districts.	c.	Technically integrated under NARSC.
			d.	Works closely with other agricultural projects in the region.

Figure 2: Primary functions and linkages of organisations within the National Agricultural Research System, Nepal (continued)

Organisation	Functions	Linkages
PAC	a. Implements extension and training programs in specific target area in the Eastern Hill region supported by an on-station and on-farm research program.	a. Supported by the British ODA.
	b. Acts as a center for basic research, foundation seed production, fruit and forest tree multiplication, and production of exotic livestock.	b. Participates in the National Crop and FSR Workshops.
	c. Serves and supports agricultural development in the Eastern Hill Districts.	c. Technically integrated under NARSC.
		d. Works closely with other agricultural projects in the region.
IAAS	a. Produces graduates in agriculture and animal science.	a. Advised and guided by Tribhuvan University, Ministry of Education.
	b. Carries out research on crop, livestock, veterinary, horticulture, social sciences, etc.	b. Linked to NARSC through RCC.
IF	a. Produces graduates in forestry.	a. Advised and guides by Tribhuvan University, Ministry of Education.
	b. Carries out research on forestry.	b. Linked to NARSC through RCC.

performance of extension services, the weakness of feedback to researchers has reduced their value as a research tool (Kayastha et al., 1989).

- The Agricultural Research and Production Project (ARPP) funded by USAID to build on the experience of the ICP by strengthening research planning, coordination and procedures, including the incorporation of systems perspectives, was established in 1985 for a five-year period. It provides support to new national bodies for research planning (NARCC), and for research coordination and procedures (NARSC). New work under NARSC is located both in newly-formed divisions (Farming Systems Research and Development Division — FSRDD and Socio-Economic Research and Extension Division — SERED) and in projects and programmes (Seed Technology Improvement Project — STIP — biofertiliser, livestock, pasture and fodder). Some achievements of these include: the FSRDD's interdisciplinary research programmes at specific FSR sites, supplemented by 'joint treks', to cater for the diversity of agro-ecological conditions, particularly in the Hills; the strengthening of research planning through quarterly meetings of the NARCC; the strengthening of research procedures through NARSC guidelines; reviews of the performance of research stations through SERED and socio-economic inputs to work at FSR sites through FSRDD; a programme over 4 Districts by STIP to stimulate production and sale of seed by small farmers; and STIP drafting of legislation on seed supply.

Difficulties

Two types of difficulty are discussed here: those specific to the new organisational efforts and approaches under NARCC and NARSC, and those of a more general nature which, nonetheless, have a bearing on the likely future performance of NARCC and NARSC.

As far as difficulties specific to NARCC and NARSC are concerned, a 1987 mid-term review of the ARPP indicated a number of constraints to progress in ARPP support of NARCC, NARSC and related divisions. It concluded, for instance, that the initial emphasis on development of FSR at headquarters level by the FSRDD had been partly misconceived: FSR is an approach for field-level work; attempts to institutionalise it at headquarters level are likely to be counterproductive by distracting attention from its proper role.

Expatriate staff responsible for FSR in the ARPP therefore had their contracts terminated ahead of schedule or rewritten.

More recent difficulties (de Boer, 1989) include delays in the provision of HMG(N) staff and services to ARPP. The agreed value of these was to have been US$3.9m over 5 years but less than half this amount had been disbursed by the fourth year of the project, necessitating accelerated disbursement of USAID contributions in order to keep the project on schedule.

By early 1989 (ie. 3.5 years into the ARP project) HMG(N) had still not made permanent the allotted number of posts in NARSC and in divisions such as FSRDD and SERED. By mid-1987, only the 10 posts designated to the NARCC had been made permanent, a further 39 remaining vacant or being filled by temporary recruits, or by officers on temporary secondment. Two years later, there had been so little improvement in the position that the prospects of a further phase of funding (from 1990) appeared seriously threatened. Whilst bearing strongly on the prospects for success of the ARPP, problems of staff deployment are not unique to ARPP, and are viewed below in the broader context of staff management in the Nepal public service.

Numerous instances have been reported of inefficiencies in ARPP programmes and projects as a result of low standards of project management. Basnayat and Yazman (1989) for instance, report that the Animal Breeding Research Division had been out of stock of Holstein-Friesian semen for over 6 months in 1988-89 despite strong demand for it by farmers in dairy artificial insemination programmes.

Broader difficulties in the national agricultural research service fall into three categories: manpower, emoluments and financial provisions.

It should be stressed at the outset that there is no shortage of trained manpower in Nepal. Data from 1983/84 indicate that over 1100 graduates were then employed in the public service in agriculture-related disciplines (Yadav, 1987). Of these, over 370 were trained to MSc and 28 to PhD level. By 1987 the number of MoA posts for first degree or higher level staff had increased to over 1400 (Freeman, 1987). First degree courses in general agriculture and animal science are offered by the Institute of Agriculture and Animal Science of Tribhuvan University. In recent years it has produced some 100 graduates per year, and has proposals in hand for an MSc course. The large number of foreign masters and doctorate-level traineeships available in donor-assisted projects means that substantial numbers of Nepalese professionals are returning annually with higher degrees. For the years 1981-82 to 1983-84, for instance, the number of long-term trainees abroad (mainly at master's level) was 46, 43 and 54 in the respective years. In addition the numbers taking short-term training were 61, 34 and 50. A brief review of staffing in the main MoA research departments suggests that there are currently over 60 PhDs against 34 in 1983 — more than enough to outweigh the anxieties expressed by Yadav (1987) that Nepal is lagging in this respect. As Kayastha et al. (1989) note, however, shortages of professional staff exist in some projects and programmes, and surpluses in others. We argue below that this has less to do with the overall supply of graduates than with management and incentive systems.

Professional posts in the public service are classed as 'non-gazetted' at the junior (ie. certificate or diploma) or 'gazetted' at the higher levels. A bachelor's degree is a minimum entry requirement for gazetted posts. Not all gazetted posts form part of the permanent civil service establishment. Freeman (1987) indicates that in the MoA in May 1987 only 66% of the 1409 gazetted posts among all three grades of professional officers were permanent, the proportion of temporary posts in second and third grades being more than twice as high as in the top grade.

Prior to 1968, the number of permanent public service posts being created in agriculture could easily absorb the number of graduates returning from abroad. Thereafter, supply began to outstrip demand, and a Public Service Commission examination was introduced. Despite the gradual expansion of research programmes under the MoA, and in separately-funded agencies such as LAC, PAC and ICIMOD, the situation to date has remained one of oversupply of graduates in relation to demand, so that those seeking permanent employment in MoA research generally first have to spend a period (generally 3-6 years) in temporary positions.

Although basic scales are identical for permanent and temporary employees, numerous other differences in conditions are likely to cause low performance among temporary staff. They include the fact that temporary posts offer no prospect of promotion and do not allow merit points to be accumulated towards any subsequent promotion. Temporary officers therefore frequently switch among posts in the search for a strong springboard to permanent employment, and are reluctant to be posted far from Kathmandu, where interviews are held. Furthermore, temporary staff are eligible for few of the benefits outlined in Table 10.

Other rigidities in personnel management likely to cause low performance include the stipulation that promotion for permanent officers is allowed only within the Faculties in which they currently serve. There are 22 Faculties in total relevant to agricultural research, so that the inability to equate demand and supply through promotion is a major source of inefficiency.

As regards emoluments, data in Table 2 suggest that excess supply of qualified staff has allowed HMG(N) to increase salaries at less than the rate of inflation. Thus, over the 23-year period 1966-67 to 1989-90, salaries of the higher-level professional grade (Class II) had risen by 232%, against a rise of 686% in the food price index. The value of allowances has also been eroded. This has particularly adverse implications for fieldwork: as Galt and Mathema (1987) have noted, per diem allowances no longer cover the basic necessities, thus threatening such innovations as the 'joint trek'.

The combination of low pay and generous leave allowances encourage the more capable staff to take on other assignments to supplement earnings. This entails the risk of a conflict of interest with their official duties. Finally, the high proportion of non-monetary benefits in overall emoluments, and the fact that both monetary and non-monetary rewards are largely unrelated to performance, means that middle and senior staff (including those in managerial positions on research stations) are unlikely to risk hard effort and unpopularity in bringing together the diverse components of staff, materials and infrastructure necessary to keep stations running effectively. Survey results reported by Yadav (1987) reveal that many research staff are unhappy with the apparently arbitrary nature of transfers and perceive little reward to the effort and rigour necessary to produce research papers.

More general financial difficulties relate to the low overall level of funding for agricultural research through HMG(N) which, prior to the inauguration of LAC and PAC, was particularly hard felt in Hill areas, the allocation being low in relation to overall population levels there, whilst the diversity of agro-ecological conditions in the Hills would require proportionately higher levels of funding than in the Tarai to achieve a similar pace of innovation. However, as Yadav (1987) notes, the delays with which agreed funds are made available has also had a particularly negative effect on performance.

An analysis of donor-funding for agricultural research and extension suggests that it has increased five-fold from the early 1970s to the mid-1980s. Even without considering the combined annual budgets of LAC and PAC of approximately US $1.5m, this means that donor resources have overtaken those of HMG(N) in financial terms, bringing overall expenditure close to the average of 0.56% of agricultural GDP for 51 countries surveyed by Oram and Bindlish

Table 2: Nepal Government salaries in relation to the food price index

	Salary (Rs/month) Gazetted — Class II	*Cumulative % increase*	*Food Price Index*
1966-67	650	—	100
1977-78	750	15	167
1979-80	885	36	205
1982-83	1095	69	283
1985-86	(B) 1625	150	353
	(A) 1825		
1987-88	(B) 1725	165	483
	(A) 1925		
1989-90	(B) 2160	232	686
	(A) 2410		

Note: From 1985-86, Class II was divided into Under-Secretary (A) and Assistant-Secretary (B) ranks.

Source: Computed by the authors from DFAMS Agricultural Marketing Information Bulletin and other sources.

(1981), and nullifying fears expressed by Yadav (1987) and Sharma and Anderson (1985) that expenditure had been stagnating at 0.22% of agricultural GDP (see Table 5).

Bolivia

Administration

The Bolivian Institute for Agricultural Technology (IBTA) was established in 1975 as a financially and administratively independent entity with responsibility for agricultural research and extension in all Departments except Santa Cruz. Overall management is vested in the central office in La Paz, whilst Departmental offices are responsible for coordinating and monitoring IBTA activities undertaken within the Departments and for coordination with decentralised development initiatives.

IBTA has a total of 12 experiment stations and 13 nurseries or demonstration plots. These are intended to generate technology for the entire range of livestock (cattle, sheep, camelids, rabbits), annual crops (coarse grains, wheat, oilseeds, grain legumes, root crops, vegetables), perennial crops (citrus, rubber, coffee, coca, chestnuts), and pastures found in Bolivia. In 1988, total research staff numbered 128, with 123 extension agents and a further 13 staff in the central office (ISNAR, 1989). Extension is managed via a total of 75 local offices which, for 1988-89, proposed over 250 demonstrations to take place in some 600 villages.

Finance

IBTA's budget from national resources has recently increased to some US$6.8m., and as discussed below, this is supplemented by foreign assistance for specific projects (IBIA, 1987). Other agricultural research and extension activities outside IBTA are also funded by donors. In 1987, approximately 85 donor-assisted projects were underway in agriculture and forestry, having an annual value of almost US $100m. The technology development component of these is impossible to assess accurately from the information available, but, as a rough estimate, it appears to comprise at least 20% of the total, making donor funding for agricultural technology development and dissemination greater than that funded by the Bolivian government by a factor of almost three.

Specific donor-funded projects in collaboration with IBTA include:

- The USAID/IBTA coca-substitution project in the Chapare
- The IFAD integrated rural development project in northern Chuquisaca
- Netherlands government assistance in fertiliser trials and promotion
- GTZ support to High Valleys agriculture in Cochabamba and La Paz Departments.

Achievements

Despite the difficulties outlined below, IBTA has made substantial improvements in agricultural technology over the last decade. These include the introduction and testing of new varieties of potato, maize, quinua, wheat, coffee, cassava and cocoa. Some success has been achieved in small ruminant improvement.

Work conducted at IBTA is organised on commodity and disciplinary lines. No serious attempts appear to have been made to introduce innovative approaches to technology development, such as farming systems research and the interdisciplinarity and rapid assessment techniques that this implies.

Difficulties

Particular difficulties for IBTA are caused by low staff salaries, which in turn derive from low budget levels in relation to the institute's overall remit. Table 3 indicates the low overall technical qualifications achieved by IBTA staff. Fewer than 6% hold a postgraduate degree, and none have PhDs, compared with approximately one-third at MSc level in Nepal, and a further 2.5% with PhDs. The relatively junior status of IBTA staff is related to the higher rate of turnover caused by low salaries and poor working conditions, discussed in more detail below.

A summary of average professional staff salaries and the rate of staff turnover is presented in Table 4. Dollar-equivalent salaries were exceptionally high in the late 1970s and early 1980s when the Bolivian peso was overvalued. In 1984-85, by contrast, local salaries were unable to keep pace with hyperinflation and the rapid depreciation of the peso, and annual turnover doubled from 12-15% of total staff to 20-30%.

Table 3: Professional level of IBTA Staff, Bolivia, 1988

	Research	*Extension*	*Project*	*Office*	*Total*
MSc	5	1	7	4	17
BSc Agronomy	21	20	12	2	55
BSc Gaduands	49	63	14	1	127
Diplomates	8	37	6	3	54
Others	5	2	1	3	11
Total	88	123	40	13	264

A recent review of agricultural research and extension in Bolivia (ISNAR, 1989) highlights a number of serious constraints to the effectiveness of IBTA:

- IBTA, in practice, does not enjoy the degree of autonomy envisaged.

Table 4: Summary of salary levels and staff turnover in IBTA, Bolivia

Year	*Average Monthly Salary[1] (US$)*	*Number of Professional Staff Resignations*	*Total Professional Staff*	*Resignations as % of Total*
1981	367.20	nd[2]	239	nd[2]
1982	56.12	22	239	9.2
1983	46.98	39	246	15.9
1984	11.61	30	246	12.2
1985	18.71	60	285	21.2
1986	45.64	74	247	30.0
1987	50.18[3]	71	252	28.2
1988	60.00	57	257	22.2

Notes: 1) Average for middle-level staff (those having completed BSc course work but not the thesis which is an integral part of the qualification)
2) Data unavailable
3) Estimate

- There has been exceptionally high turnover in the number of Executive Directors in the last 3 years, resulting in lack of cohesion and discontinuity.
- Low overall budgets, delayed authorisation of disbursements, low staff salaries and delayed payments have led to high turnover and low morale. Low operating budgets and cumbersome procedures mean that equipment cannot easily be repaired or replaced, that consumables are in short supply and that field visit programmes are curtailed.
- Research/extension links are exceptionally weak at both field and headquarters level. The research and extension subdirectorates draw up forward plans which are inadequately coordinated. Collaboration at the field level is made particularly difficult by the remoteness of many extension offices and poor communications.
- Particular problems are posed for the conventional centrally controlled extension service by the dispersed patterns of settlement, poor communications and wide variations in the local importance of crops or livestock. IBTA has estimated that only 2% of farmers have any contact with extension services. Yet, at the same time, no mechanism has been developed (except in Santa Cruz) to involve the numerous agricultural programmes operated by NGOs, national or foreign-assisted development projects and others into technology dissemination and feedback.

- Inadequate resources, low motivation and staffing instability are part of a vicious circle in which forward plans are drawn up without conviction, and then remain largely unfulfilled, so bringing discredit to the institute.
- A tendency for staff at all levels to take on additional assignments which encroach on their office time, in order to supplement low incomes.

Agricultural research and extension in Bolivia and Nepal: a comparative summary

Based on the foregoing discussion, some of the principal features of the national agriculture technology development systems in Bolivia and Nepal are summarised in Table 5.

Whilst the data must be regarded as tentative, the orders of magnitude are sufficiently clear to allow some general conclusions to be drawn:

- Overall national expenditures on agricultural research are similar between the two countries, but higher in Bolivia per head of rural population.
- Staff levels are similar in relation to the size of rural population.
- Much of Bolivia's additional non-staff expenditure is likely to be accounted for by the difficulties of maintaining even a skeletal

Table 5: Main quantitative indicators of national research systems in Bolivia and Nepal

	Bolivia	*Nepal*
Graduates employed in research and extension[2]		
— overall	200	1400
— per/m rural population	62	89
National budget for research and extension		
— overall (US$ m)	6.8	2.0
— $m per/m rural population	2.0	0.1
National budget plus foreign assistance for research and extension		
— overall (US$ m) (approx.)	25.0	8.0
— $m per/m rural population	7.8	0.5

Notes: 1) All data should be treated as approximate.
2) For Bolivia, includes graduands, but not diplomates.
For Nepal, data for both manpower and financial resource allocations for extension are incomplete. In both cases, all estimates exclude the resources committed through the case studies presented in this book.

research and extension service in countries where terrain is difficult and population density low (average 6 persons/km^2 in Bolivia against 120/km^2 in Nepal).

Foreign funding for research and extension is much higher in Bolivia than in Nepal, principally as a result of donor concern to identify alternatives to cocoa cultivation.

urther comparisons suggest:

that low salaries, poor motivation and 'moonlighting' are major difficulties in each case, the problems of high staff turnover being prevalent in Bolivia and reluctance to make appointments permanent prevalent in Nepal.

the much lower proportion of staff having postgraduate qualifications in Bolivia is, no doubt, related to the problem of high turnover, but may also be influenced by the higher availability of opportunities in, for instance, the commercial agriculture sector and in NGOs.

3
Alternative Approaches to Agricultural Research for Complex, Diverse and Risk-Prone Areas in Bolivia and Nepal

Introduction

In the opening Chapter, a case was argued in general terms for increased resource allocations to difficult areas and for the development of institutional forms and research methods appropriate to the management of such resources. We now turn to the specific factors underlying the introduction of innovative approaches in Nepal and Bolivia.

The **first** is a growing awareness that a centrally administered system tends to favour the better-endowed areas. It was noted above for Nepal that the bulk of agricultural research resources had been allocated to the Tarai and Kathmandu valley (Map 2). Evidence presented in Annex 1 on improved varieties of rice, maize and wheat indicates that most are recommended for lowland conditions. As discussed below, adoption levels are low even for those varieties specifically recommended for the Hills. Livestock is particularly important in the Hills, yet as Sharma and Anderson (1985) note, research expenditure on this is low in relation to its contribution to GDP. Much livestock research has focused on milk production among cattle, to the neglect of buffalo and small ruminants which make up the bulk of Hill livestock. Expenditure on major cereal crops is high, some 70% of biological research experiments being conducted on these (Sharma and Anderson, 1985), yet, given the conditions of the Hills, both crop and livestock research are characterised by major methodological shortcomings insofar as they tend to be conducted on a single commodity basis with little reference to systems interactions. Animal manure in the Hills is of major importance, as is fodder from crop residues or (frequently off-farm) trees. Such trees are also a valuable source of green manure. Yet, the desirable features of by-products tend to be overlooked in commodity approaches, and the generation of off-farm biomass is neglected altogether. In Bolivia, research by the NARS has focused on crop and animal technologies relevant to the high Andes and Valleys to the north of La Paz, with some emphasis also on tree crops for lowland areas further to the north. Initiatives to develop technologies for either mechanised or manual cultivation of the lowlands of eastern Bolivia were, until the establishment of CIAT in 1976, extremely limited.

The **second** is a recognition that even the limited resources allocated to difficult areas were poorly managed under a centrally administered system: disbursements from the administrative centre to outlying areas tended to be delayed; inadequate delegation of authority to the periphery caused delayed decisions; poor communications with central stores and repair facilities added to delay; allowances for living on remote stations tended to be inadequate, thereby causing discontent among staff, or contributing to low attendance and poor timekeeping resulting from lengthy commuting from more desirable residential areas.

The **third** is growing recognition of the importance of these difficult areas within the political economy. In Nepal, widening divergences in living standards between Hill areas on the one hand, and Tarai/Kathmandu Valley on the other, are likely to lead to unmanageable internal migration, and increasing discontent among the high proportion of population resident in the Hill areas.

Improved technology for the Hills is also necessary on production grounds: official statistics showing a declining trend in per-hectare yields, are known to be unreliable (Balogun, in preparation) yet there remains widespread concern that the productive base of Hill farming systems is threatened as population density increases. Certainly, there is evidence that such important resources as forest are in decline, both qualitatively and quantitatively (LAC, 1989). In Bolivia, the recognition that more had to be done for technology development in the eastern lowlands was motivated by different factors. The central government began to recognise the area's potential for large-scale mechanised agriculture in the 1950s, but was slow to commit adequate development resources to the area. The situation changed as, first, Development Corporations having a high degree of decentralised authority were set up in each of the country's eight Departments in the 1960s and, then, oil and gas were exploited in the eastern and southern lowlands from the late 1960s onwards, giving (especially) the Santa Cruz Development Corporation sufficient resources to finance agricultural research. This was initially linked to ambitious proposals for agro-industrial development in the Department, but from the late 1970s increasingly also supported small non-mechanised producers. Secondly, in the case of Bolivia, even in the 1950s, central government had seen the potential of the eastern lowlands as an area of settlement for small farmers wishing to move from the declining resource base of the Altiplano, but was slow to provide necessary infrastructure and support services. The collapse of tin mining in the Altiplano stimulated further internal migration in the 1970s and 1980s and central government responded with infrastructure to facilitate settlement, though such efforts were poorly organised and underfunded, and the majority of families have settled spontaneously rather than in these 'managed' settlements.

A final set of factors now relevant to technology development in the eastern lowlands, though not influential in the decision to establish CIAT or in its early years of operation, is the increasing realisation that improved management practices and technologies are required in both mechanised and non-mechanised farming if they are to continue in a sustainable fashion.

We now turn to a consideration of the case studies of institutional innovation in Nepal and Bolivia

The physical setting

The contrast between the environments of eastern Bolivia and the mid-Hills of Nepal could hardly be greater: in Bolivia, the Department of Santa Cruz served by the Research centre for Tropical Agriculture (Centro de Investigación Agricola Tropical — CIAT) and the British Tropical Agriculture Mission (BTAM) covers some 370,000 km^2 (Map 1), with a population of approximately 1m persons, constituting approximately one-third of the national territory and one-fifth of national population. More than half of the Santa Cruz Department population live in urban areas, so that rural population density averages roughly 1 person/km^2. Most of the Department is rolling plain of approximately 300 m altitude, which spans the fringes of both the Amazon and Paraguay river basins. In the west and south of the Department, the Andean foothills rise to some 2000 m. Mean annual temperature is 24.5°C with an annual rainfall ranging from 2300mm in the west, to 1100mm for much of the remainder of the Department, except for drier areas in the South.

The mid-Hills of Nepal — eleven Districts in the East being served by Pakhribas Agriculture Centre (PAC) and a further eleven Districts in the West and Centre by Lumle Agriculture Centre (LAC) — are characterised by sharply dissected terrain, much of the farm land lying in the range 500 — 3000m (Map 2). Mean annual rainfall varies between roughly 500mm and 2000mm with some 70% of precipitation between June and September. The mid-Hills cover 62000 km^2 (ie. 42% of national territory and 47% of national population). Two-thirds of the mid-Hills area is now the responsibility of LAC and PAC.

Population density is over 70 persons/km^2, but the density in relation to cultivated land — at 870 persons/km^2 is among the highest in the world.

Communications are poor in both locations: in eastern Bolivia, railways connect Santa Cruz City to Brazil and to Argentina, and the Department's two main stretches of surfaced road connect Santa Cruz with the remainder of Bolivia to the west. A surfaced road penetrates the eastern mid-Hills of Nepal into the PAC area north of Dhankuta but all transport to and from the road is by pack animal or headload. Construction of a road into the LAC area is underway but, again, feeder roads do not exist. In both countries, telecommunications are extremely limited, relying principally on two-way radio, and postal services are poorly developed.

Farming Systems

Farms are much more diverse in type and size in eastern Bolivia than in the mid-Hills of Nepal. Table 6 presents data assembled in the mid-1980s on farm size, principal enterprises and gross farm income for five diverse groups of farmers in the Santa Cruz Department. Striking features include the importance of commercial farming in arable and livestock enterprises, the diverse ethnic origins of farmers and the large smallholder sector. Rice grown under slash and burn cultivation is the principal subsistence crop for smallholders, maize and sugar cane production being important in some areas.

The mid-Hills of Nepal are more uniform in terms of overall farm size (most falling in the range 0.25 — 0.4 ha) and technology, most farmers using human labour and animal draught, with very limited quantities of external inputs. Table 7 presents HMG(N)'s official data on the production of principal crops in the mid-Hills, though it should be noted (Balogun, in preparation) that survey methodology has been poor, so that the data (particularly those indicating changes in yield) are highly unreliable. Other major contrasts with eastern Bolivia include:

- the absence of unoccupied land for cultivation in Nepal; settlement patterns have a long history and complex social structure; the techniques that have evolved to make arable farming sustainable on infertile soils and hilly topology require substantial labour to maintain terraces and incorporate organic matter.
  By contrast, large-scale settlement in eastern Bolivia began only in the 1950s and, until recently, farming practices were universally exploitative, ignoring resource maintenance requirements.
- cropping patterns are spatially and temporally complex in Nepal. Maize/millet intercrops are common, and LAC technical reports (eg. Sthapit et al., 1988) detail over 20 common cropping sequences within the agricultural year, determined mainly by altitude and moisture availability. In Bolivia, whilst some rotations designed to maintain fertility have long been practised (such as maize + soya among the Mennonites), it is only through recent research and dissemination efforts that sustainable practices integrating crops, trees and animals are being introduced
- in Nepal, critically important biomass, especially fodder and green manure, is derived from off-farm sources such as isolated trees, verges and woodland. These resources fall outside the remit of conventional on-farm research, and particular efforts have to be made to cater for them in technology development programmes
- the particularly difficult communications in the mid-hills of Nepal mean that for many areas the export of labour, either to more developed parts of Nepal or abroad, is more profitable than the sale of agricultural produce. Adult males are therefore absent from many households for long periods, and remittances play an important role in the household economy (Seeley, 1989).

The institutional setting

Institutional issues have been discussed in depth in Chapter 2. Only the principal characteristics necessary to an understanding of the broad context are presented here. An important feature common to both locations is that the institutions enjoy a high degree of independence from central government. In Bolivia, the central government is represented on the governing body of CIAT, and provides some funding, but CIAT is primarily responsible to, and funded by, entities within Santa Cruz Department, particularly the Departmental Development Corporation. In Nepal, LAC and PAC have maintained financial and administrative independence whilst operating within a broad remit provided by HMG(N), though moves are now underway to incorporate them into the HMG(N) agricultural research system.

Both CIAT and LAC/PAC were established in the last 15-20 years. LAC and PAC are unusual in that an agricultural research and extension capacity based largely on local staff has been developed within what have essentially been UK-funded and managed institutions. CIAT, by contrast, is Bolivian in its design, management and funding, but since its inception has received technical support from a team of British advisors (and, at various times has received advice in specific issues from other foreign-assisted programmes).

Case Studies from Nepal: Lumle Agriculture Centre (LAC) and Pakhribas Agriculture Centre (PAC)

Departing from the premise argued above, namely that a case does exist for identifying approaches to the organisation and management of agricultural research to address the specific conditions of the Hills, we now examine efforts to do so that have been made over almost fifteen years, seeking to identify whether these approaches might be more widely applicable in the Hills and whether they offer more general lessons for agricultural research in difficult environments elsewhere. The physical features of LAC and PAC command areas are outlined in Annex 2.

Following a brief historical review, we discuss the current status and evolution of experience at LAC and PAC in respect of:

- research philosophy and organisation
- research methods
- progress in technology generation and the contribution of exotic genetic material
- financial and personnel management
- integration with the HMG(N) research system

LAC is located in the Western Hills of Nepal, 20 km west of Pokhara in Kaski District. It was established in 1968 at the request of HMG(N) to train retired Gurkha soldiers in improved hill farming practices on their retirement from the British Army. PAC, situated in the Eastern Hills in Dhankuta District, approximately 25 km

Table 6: Santa Cruz Department, characteristics of the principal groups of farmers (1982-84))

Group	Number of farmers	Average area cult./ livestock kept	Average tractors per 1000 ha	Mean GFI (US$)	Enterprise type and % of GFI	
Commercial farmers						
Camba	1000	66 ha 73 layers 35 dairy cattle	39	23790	Cane Chicken Cotton Soya Maize Eggs Milk Total	45 14 12 12 8 6 3 100
Japanese	432	46 ha 1050 layers	20	33170	Eggs Chicken Rice Soya Cotton Total	69 13 12 4 2 100
Mennonites	2025	18 ha	38	3960	Soya Maize Eggs Rice Total	65 26 7 2 100
Semi-Commercial Famers						
Camba ranchers	2000	400 head	—	10140	Cattle	100
Peasant farmers						
Colla	20100	4.7 ha	8	380	Rice Cane Maize Milk Total	57 36 6 1 100

Notes: 1. The mean annual value of sales of each commodity (based on factory gate prices), averaged over 1982-84 is calculated for each group. These are then aggregated for the commodity sets relevant to each group and divided by the number of farmers in column 2 to provide an estimate of GFI per farm.

2. No estimates were available for the value of production for Mennonite milk (which is transformed into cheese). An informed guess might be that the value of Mennonite milk production is approximately equal to that of liquid milk sales, raising their GFI by about US$410.

Source: Factory gate prices from the Agricultural Marketing Department of CORDECRUZ. Production figures from CAO. Information on tractor use from CAO and Tejada.

northwest of Dhankuta, was established in 1973 with the same objective.

A chronology of major events at each location is given in Annex 3. Briefly, following the review by the Rockefeller Foundation team in 1975 which found that many technologies developed in Nepal were unsuitable for adoption by farmers because they had been developed in isolation from farmers' conditions, HMG(N) requested LAC and PAC to broaden their mandates to include the development of agricultural technology, and to cater to all farmers in surrounding areas, not only to retired servicemen.

The UK Overseas Development Administration has been the sole funding agency for LAC and PAC since then, and is currently considering proposals for continued funding with contributions from HMG(N) to 1995.

As their research capacity built up, both developed outreach areas in which close linkages could be fostered with farmers in the development of technology. These took different paths, partly reflecting minor differences in philosophy, partly specific agro-ecological conditions. PAC's outreach responsibilities were increased in 1982 when it was requested to provide research support for the four Districts of the Koshi Hills Agricultural and Rural Development Project (KHARDEP). In 1986, LAC's outreach was expanded to the four districts of the Hill Food Production Project. Both were functionally incorporated into NARSC in 1988 and were recognised as research centres for their respective development zones within the national network of regional research farms and stations, each now being requested to service eleven Districts, the total of 22 Districts being almost half of those falling outside the Tarai and Kathmandu Valley.

Philosophy and organisation

There has been a gradual evolution of philosophy at LAC and PAC as each have responded to local needs and opportunities.

Initial activities at LAC included farmer training at the Centre, extension activities and the provision of inputs (cereal and horticultural seed multiplication, fruit and fodder trees, livestock). Strong technical support from LAC's researchers to its own extension staff was premised on the need to respond to farmers' requirements and opportunities from the very beginning.

Developments at PAC in the early period were similar but with two major differences:

Table 7: Area planted, production and yields for the mid-hills of Nepal (1975-1985)

	Area Planted '000 ha			*Total Production '000 metric tons*			*Yield (kg/ha)*		
	1975	*1985*	*% change*	*1975*	*1985*	*% change*	*1975*	*1985*	*% change*
MID-HILLS:									
Rice	195.8	288.3	47.2	489.6	581.2	18.7	2501	2016	–19.4
Maize	275.2	384.1	39.6	506.5	523.5	3.4	1840	1363	–25.9
Wheat	19.6	167.8	83.2	101.4	193.5	90.8	1107	1153	4.2
Millet	88.8	102.8	15.8	104	94.2	–9.4	1171	916	–21.8
Barley	10.4	13.3	27.9	10.2	11	7.8	981	827	–15.7
NEPAL:									
Rice	1240.9	1367.4	10.2	2491.0	2756.9	–10.7	2007	2016	.4
Maize	454.7	565.7	24.4	796.1	818.3	2.8	1751	1447	–17.4
Wheat	297.8	468.9	57.5	342.0	588.5	72.1	1148	1255	9.3
Millet	125.3	136.3	8.8	142.1	125.7	–11.5	1134	922	–18.7
Barley	27.2	27.3	.4	25.2	23.1	–8.3	926	846	–8.7

first, production and distribution of new material became important only in the early 1980s;
second, at LAC a separate extension section was established only in 1987, prior to which each technical section (agronomy, horticulture, livestock, forestry) was responsible for organising its own extension activities, whereas crop extension at PAC was covered by a separate extension section from the beginning.

Thinking at the two centres evolved from this early experience in remarkably similar ways. The basic tenets of the approach were:

- That, through a combination of on-station and on-farm trials, a capacity should evolve which would both be responsive to farmers' needs and opportunities, and would take advantage of new technologies becoming available in a wider arena, both nationally and internationally.
- That transport problems and the limited resources of government institutions in outreach Districts made the supply of inputs precarious. Priority should therefore be given to technologies which do not rely heavily on external inputs. At the same time, the production and supply of inputs (seed, other planting material, animal stock) is an issue in which the research centres need to take a leading role, with the eventual objective of handing over a large part of the responsibility to farmers themselves. Research and extension staff would need to collaborate closely in both technology testing and in input supply.
- That farming systems are complex so that commodity or discipline-based approaches, whilst important in some contexts, cannot address field issues fully. Of particular concern at both locations are first, the high degree of interdependence among components of the farming system. The importance of crop residues and thinnings as animal feed are widely known, and have influenced farmers' varietal selection and planting densities. Second, much energy is derived from off-farm natural resources, as evidenced by the long hours spent (chiefly by women) in the collection of firewood and fodder (Seeley, 1989). The importance of these interactions lies not only in the short-term benefits derived, but also in their long-term contribution to the sustainability of Hill farming systems. Without animal manure, for instance, it is estimated that soil fertility and texture would decline to levels likely to make arable farming uneconomic (Gibbon, pers. comm.). The maintenance and, where possible, expansion of key

on- and off-farm resources such as livestock and forest therefore merits substantial allocations of research effort and the development of research approaches over and above those linked to individual commodities or disciplines. LAC has responded to these requirements by establishing four interdisciplinary research 'thrusts', in fodder, soil fertility, income generation and genetic improvement, the first two of which focus particularly on issues of longer-term resource sustainability normally ignored by commodity or disciplinary approaches. In the same way, fodder and soil fertility are two of the seven interdisciplinary 'working groups' defined by PAC. Further evidence of the importance attached to the identification of appropriate research methods and priorities is provided by the proposal that, as expatriate staff are reduced over the period 1990-95 to facilitate the integration of LAC and PAC into the HMG(N) system, the sole remaining expatriate at the end of his period will be a research adviser (at PAC, specifically a *farming systems* research adviser).

The structures which evolved at LAC and PAC to implement these approaches differ somewhat:

At LAC the head of each section is responsible for the on and off farm activities of that section, and all staff are involved in both types of activity. The sections design on-farm trials using material obtained from national programmes (and, prior to integration into NARSC, also occasionally directly from abroad). The trials are then implemented by LAC extension staff in the Extension Command Area for agronomy and horticulture sections, but monitored by the sections themselves. Forestry and livestock sections' trials are implemented by their own field assistants. Apart from individual trials led by each section at various points in the ECA, there are also three Farming Systems Sites (each one consisting of a village) at which trials known to have strong systems interactions are planned jointly by all sections and monitored by site-based staff under the socio-economics section. Section staff also participate in 'thrust' group trials. These are explicitly multidisciplinary trials in three broad areas: fodder, soil fertility and income generation. Thrust activities are reported at monthly technical meetings for all staff. With the full implementation of farming systems approaches, LAC has agreed that all sections should spend 60%-70% of their time in on-farm research, except for livestock where the proportion is 40%.

At PAC, two sub-centres have been opened to widen the altitude range at which agronomy research is conducted. Prior to 1982, on-farm testing took place principally in the Local and Northern Target Areas. With the subsequent responsibility for KHARDEP, the range of testing was extended to embrace these additional four Districts.

Research methods

LAC has been responsible for devising a major innovation in farming systems research methodology. In the second half of 1983 a review of adoption of LAC technologies instigated by a recently-arrived Director indicated that, although technically viable, few were being adopted by farmers. At the same time, publication of the results of multidisciplinary work conducted under the ICP/CSP made LAC staff realise that there had not been sufficient coordination among sections to produce technologies that responded to farmers' opportunities and needs. In mid-1984 the heads of all sections therefore accompanied the Director on a 10-day 'joint trek' which resulted in the identification of farming systems sites in two of the villages visited. A subsequent joint trek was used to identify multidisciplinary research priorities at two further locations. The joint trek — subsequently renamed 'Samuhik Bhraman' — has subsequently been adopted by both PAC and FSRDD. Joint treks continue to be used in LAC to visit FSR sites at least twice yearly, and to visit the trials of individual sections and of thrust groups within the Extension Command Area (ECA).

Methodologies developed in the ICP/CSP were implemented with appropriate local modification by PAC in 1983, prior to the LAC's development of the Samuhik Bhraman. These were used to identify representative research sites within the four Districts of KHARDEP. As at LAC, the socio-economics section was heavily involved in providing site descriptions of these, and in evaluating the adoption of PAC technologies. In this way, the focus of the technical research programmes is being sharpened, and the recommendation domains being redefined on a continuing basis. As at LAC, on-farm trials are designed for three altitude ranges and three land types (upland; lowland irrigated; lowland rainfed) using common cropping patterns. Details of the predominant cropping patterns at both LAC and PAC are given in Annex 4.

PAC has developed the concept of 'working groups', analogous to LAC's 'thrust groups' to broaden the scope of its on-farm work from cropping systems to a full farming systems approach. The seven working groups include: farming systems; soil fertility, conservation, management and improvement; fodder production, management and utilisation; information dissemination; women's participation in development; research and extension methodologies, and sustainable low input technologies.

As far as specific trials methodologies are concerned, both LAC and PAC collaborate with national programmes in evaluating new technologies (principally crop varieties) through farmers' field trials. In addition, minikits and preproduction verification trials (PPVT) use material from both the Centres and from national programmes. PAC conducts a similar range of trials, in 1987 having conducted 99 FFTs and 10 PPVTs. PAC also distributed 450 summer minikits (rice, maize and millet) and 356 winter minikits (wheat, achieving a feedback of over 50%, much higher than that obtained in the national programme).

PAC has made two important adaptations to research methodology which enhance overall efficiency and make the methods better suited to local conditions. First, the minikits produced with its own materials contain two varieties rather than the one contained in national programmes. The farmer makes a three-way comparison — among each of these and his own variety — and then reports it back to PAC, so becoming more of a partner in research. Second, the initial concept of PPVTs introduced by the CSP was to test a new variety and corresponding new set of practices as a package against a local variety grown under conventional practices. PAC splits the PPVT package into two components, the new variety with and without new management practices, so that at little extra cost the amount of information from a $125m^2$ test plot is greatly increased, allowing the performance of new varieties under less than ideal husbandry practices to be assessed.

The management of field trials varies between LAC and PAC, partly because of slight differences in approach to on-farm research, partly because of differences in the type and size of outreach areas. At LAC, each of the FSR sites is staffed by a Supervisor, ie. a junior member of the Socio-Economics Section who lives at the site and monitors the trials. A locally-recruited farmer assists in monitoring and recording. At each FSR site 15-20 trials per year are conducted. The supervisor reports to monthly technical meetings at LAC and is visited on-site several times during the season by senior technical staff. Extension section staff have the responsibility of monitoring other individual sections' trials in the ECA and report in a similar way.

The situation at PAC is more complex. Trials are conducted in each of the four KHARDEP districts, and a field team is located in each, comprising an agronomist, superintendent (Junior Technician) and three field assistants (Junior Technical Assistants). The entire programme is coordinated by the Agronomy Section at PAC. The District agronomists are based in the respective District Agricultural Development Offices (ADO) and participate in ADO activities, including monthly meetings with extension staff, and training courses, when their on farm trial duties permit. On-farm trials are conducted at three sites per District at the Agricultural Service Centres (ASCs), an assistant being located at each for the duration of the trial, and visited by the District agronomist 3-4 times per season. Field assistants are usually local farmers from each site, who receive multidisciplinary research-oriented training at PAC. Field assistants are then responsible for the selection of collaborating farmers, for monitoring and recording, and for reporting to PAC on fortnightly and monthly schedules. The PAC Agronomist reports to a monthly technical meeting of all section heads on the basis of these reports and the trek (ie. site visit) reports received from the District Agronomist. The latter is not required to attend the meeting on a regular basis.

Progress in technology generation and the contribution of exotic genetic material

Ideally, in order to obtain an overview of progress across the spectrum of technology improvement, it is necessary to make an inventory of farmers' crop and livestock enterprises, the importance of each according to some uniform criterion (such as contribution to gross farm income), and the importance of specific types of innovation adopted within each of these enterprises. More details could be added by including (especially in Nepal) increases in productivity (or a reduced rate of productivity decline) in the resources necessary to sustain these enterprises, such as soils and tree products. In both Nepal and Bolivia some consideration of the longer term contribution of improved technology to the wider resource base — ie. protection of the environment, though not easily reduced to uniform criteria — is also desirable.

In practice, data are not available to permit a fully quantified assessment of this kind. In the discussion that follows, a quantitative approximation is made of the extent of adoption of certain technologies that have been developed or introduced by LAC, PAC and BTAM and have formed a principal focus of their work (viz. the introduction of new genetic material in crops and animals). In other cases, particularly in forestry and environmental protection, the accounts are more impressionistic. Where possible, the attempt has also been made to estimate the contribution made to new technology (here, specifically, genetic material) by the international research centres affiliated to the CGIAR. Such an evaluation is necessary if the prospects for success of pressures within the CGIAR to place more emphasis on technologies for poor farmers are to be assessed.

Table 8 provides an overview of the proportion of cropped area under improved varieties in each of the three case studies. An estimate is also made of the contribution to these made by genetic material from the CGIAR centres. Similar estimates are for animals in the same Table. Some caveats are in order: the data presented here are extracted from detailed Tables in Annex 5. Data on overall planted areas in these are not entirely reliable as Balogun (in preparation) indicates. Nevertheless, data from the LAC and PAC research command areas have to some extent been verified by researchers' familiarity with these areas, and so are likely to be more reliable than on average. Estimates of the proportion of area under each crop planted to new varieties were obtained from discussions with scientists at the respective research centres, as were those of the relative importance of each improved variety within this total. Charts illustrating the parentage of individual varieties were used in conjunction with scientists to assess the proportion of genetic material deriving from the CGIAR centres. This inevitably involves approximations and has the added disadvantage of attributing equal weight to all components of the genetic make-up of plants. In reality, as can be seen from varieties of wheat and rice produced in the Green Revolution, certain genes (for dwarfing and fertiliser response) were of crucial importance. A similar methodology was used for animals. The errors likely to arise from this methodology are, however, diminished by the facts that many of the improved varieties or types contain no genetic material at all from the CGIAR centres, so that the possibilities of under or over representation do not arise.

The data suggest that, while a high percentage of the land in the LAC command area under certain staple crops (maize, rice) is planted to improved varieties, in others (millet) — it is minimal. In all cases except for wheat, the contribution of CGIAR institutes' genetic material to the varietal improvement has been very small. This is not for lack of contact between LAC and these institutes. Collaborative trials with IRRI have, for instance, been underway for a decade, but the conspicuously poor performance of IRRI material has led to recommendations by a consultant to LAC and PAC that the scope of such contribution should be severely reduced and that crop improvement strategies should focus on the search for superior local material and its distribution among farmers, a recommendation which has now been adopted.

Financial and personnel management

Independence from the HMG(N) public service has allowed LAC and PAC to operate outside the financial restrictions that normally apply. Agricultural research under HMG(N) tends to receive capital account items under donor funding, so that equipment and vehicles are not generally in short supply. The principal constraints appear to lie in two main areas of recurrent expenditure: the operating costs of research stations and related services (materials, fuel, repairs to equipment) and in allowances (eg. per diem travel expenses) and remuneration to staff. Any recurrent costs requiring foreign currency are particularly subject to restriction and delay.

On operating costs, among many examples, Basnayat and Yazman (1988), report that the Animal Breeding Research Division was out of stock of Holstein-Friesian semen for its Artificial Insemination programmes for over six months in 1988-89; SERED's reports on the effectiveness and performance of agriculture research farms and stations (eg. K.C. et al., 1987 and Sharma et al., 1988, on Kabre and Surkhet respectively) detail a number of shortcomings, including delays in the release of annual budgets from central headquarters, the poor state of repair of buildings and equipment; and inefficiencies through poor communications with other organisations, with which the stations are required to collaborate.

In respect of manpower, the SERED reports catalogue such difficulties as unfilled vacancies, rapid staff turnover and the allocation of individuals into duties inappropriate to their qualifications. Such constraints inhibit the development of improved methodology for research with small farmers: Galt and Mathema (1987), for instance, note that HMG(N)'s adoption of the Samuhik Bhraman (joint trek) is, in practice, severely constrained by the fact

Table 8: Extent of introduction of genetic material in improved crops and animals in research command area of LAC, Nepal

	Rice	*Maize*	*Wheat*	*Millet*	*Barley*	*Potatoes*	*Oilseeds*	*Grain legumes*	*Other*	*Total*
LAC Cropped area/yr (ha)	64760	60000	8694	184200	1200	1500	1000	2000	323	354
% of area to improved varieties	40	55	100	5	45	40	0	20	24%	(78334 ha)
Area-equivalent of genetic material from CGIAR centres (ha/yr)	1994	660	3695	0	0	30	?	?	2%	(6379)

that per diem fieldwork allowances do not cover even modest living costs. Numerous evaluations (eg. K.C. et al., 1987; Sharma et al. 1988; APROSC, undated) have drawn attention to the shortcomings in current performance and limitations on the capacity to innovate imposed by financial difficulties of this kind. These include: limited contact with farmers; infrequent visits to trial locations; inadequate analysis and writing up of experiments; inability to expand seed production programmes in response to farmers' demands; lack of transport leading to untimely distribution and spoilage of such inputs as seed; inadequate personnel to monitor and record farmers' response to eg. minikit programmes.

Since their inception, LAC and PAC have been fully funded by the British Overseas Development Administration on both capital and current accounts. In recent years, funds have been allocated on the basis of multi-year agreements. PAC, for example, was funded from 1985-88 under a detailed ODA Projects Evaluation Committee (PEC) agreement containing year-by-year allocations under capital and recurrent subheads. This was subsequently extended by one year in 1988-89, and a request for a further six years' funding (1989/90 to 1994/5) is currently being considered. Disbursements within these periods are conditional upon satisfactory reports by monitoring and evaluation reports. HMG(N) is now making a contribution to LAC and PAC funding in the form of eg. materials for trials, land and some local staff salaries. Estimated at some £30,000 for 1989-95, this amounts to less than 1% of overall project costs.

It has already been noted that HMG(N) staff salaries and allowances are so low as to be impairing the efficiency of research, and have failed to keep pace with inflation. In Table 9 comparisons are made between basic salaries in HMG(N) and at PAC and LAC for a given grade since 1968-69.

LAC/PAC salaries exceeded those of HMG(N) by a factor of 1.5 in the early period, rising gradually to a factor of 3 by 1988-89. Annual increments for LAC/PAC salaries are made annually in response to changes in local living costs, against the interval of 2-3 years between HMG(N) adjustments.

Table 10 summaries the benefits available to HMG(N) and LAC/PAC staff of comparable grades over and above basic salary. HMG(N) basic salaries are raised by 25%-100% to provide a remote area allowance, whereas the (already much higher) LAC/PAC basic salaries are intended to cover the cost and inconvenience of living in remote areas. In most categories other than this, the allowances paid by LAC/PAC are greatly superior to those of HMG(N), the differentials having been maintained since the early 1970s.

Integration with the Nepal research system

From their establishment in 1968 and 1972 respectively, up to 1985, the development of LAC and PAC has been broadly in response to requests for technical cooperation from HMG(N) to the British Government. Some links were established with trials programmes and, from 1982 to 1989 PAC was required to include the KHARDEP area as part of its extension domain, but the two stations have remained financially and administratively independent of HMG(N).

Table 9: Comparison of salaries of middle-level professionals between Nepal Government (Gazetted Class II) and LAC/PAC employees (Sectional Head)

Years	*Salaries*			*Annual Increment*		*Ratio of*
	HMG (Rs)		*LAC/PAC (Rs)*	*HMG (Rs)*	*LAC/PAC (Rs)*	*/PAC salaries*
1968/69	650.0		990	20.0	20.0	1.5
1973/74	750.0		1540	20.0	—	2.1
1978/79	945.0		2305	25.0	104.0	2.4
1981/82	1095.0		2768	25.0	124.0	2.5
			(Max 3636)			
1984/85	1625.0	(B)*	3391	50.0	156.0	2.1
	1825.0	(A)*	(Max 4483)	50.0		
1986/87	1725.0	(B)	4377	50.0	149.0	2.5
	1925.0	(A)	(Max 5432)			
1988/89	2160.0	(B)	6542	50.0	380.0	3.0
	2410.0	(A)	(Max 9199)			

*Figures in parentheses are levels. B = Assist. Secretary A = Under Secretary.
Source: Ministry of Finance, Nepal
Lumlke and Pakhribas Agricultural Centres, Nepal

A desire for closer integration with HMG(N) research and extension services began to emerge in the 1980s, a joint Nepal/UK review mission recommending in 1982 that PAC should be designated an HMG(N) research and development station and that a major part of its extension activities be handed over to HMG(N) so as to allow the centre to concentrate on research. The 1985-88 phase of UK funding was designed to accommodate these recommendations but, in practice, the incorporation of LAC and PAC into the national agricultural research programme through NARSC was not achieved to the degree intended, partly through unresolved difficulties within NARSC (see Chapter 2). In 1988, LAC and PAC were functionally incorporated into NARSC, and from 1989, both have been recognised as regional stations for crop, livestock and forestry research within the HMG(N) research hierarchy, and each has been allocated research responsibility for eleven Districts.

It is difficult to predict whether the incorporation of LAC and PAC into NARSC will, on balance, generate positive net benefits for agricultural research and extension in the Hills. Certainly, the two centres have already made substantial contributions to methodology in the HMG(N) system, including the national adoption of the interdisciplinary 'joint trek'. It is conceivable that integration might promote the adoption of other improvements to methodology, such as the PAC approach to agricultural extension (Thapa et al., 1988) or the LAC initiatives on local seed production and supply (LAC, 1989). It also appears likely that integration into the HMG(N) system will assist the rapid spread of specific technologies such as cold-tolerant rice varieties identified at LAC and pigs attractive to low-income households bred at PAC (Gatenby and Shrestha, 1989).

LAC and PAC have long collaborated with the national system in aspects of crop genetic improvement. For instance, in 1977 the National Maize Development Programme collected a large number of maize lines from local and international (especially CIMMYT) sources. PAC's collaboration in a breeding programme allowed two crops per year to be produced — one at PAC and one in the Tarai, thus accelerating the process from which the variety Manakamana-1 emerged, and this has now been released at national level. More comprehensive links between LAC/PAC and national programmes are intended to capture wider benefits of this kind. In the proposals for PAC funding for 1989-95, for instance, PAC agrees to send representatives to a total of eight national level committees and working groups.

The more crucial question, however, is how LAC's and PAC's capacity to develop technology for the Hills will be affected by the compliance with central management norms that will be required of them in the future. As far as the introduction and testing of new technology are concerned, two effects are likely, both potentially negative: first LAC and PAC will be required to adhere to HMG(N) procedures, such as those for importing plant germplasm. This may involve both delay and the diversion of resources to argue the case

Table 10: Comparison of perquisites and benefits of middle-level professionals between Nepal Government (Gazetted Class-II Officers) and LAC/PAC (Sectional Head, Local Employees).

Particulars	HMG	LAC/PAC
1. Remote Area[a] allowance	25-100% additions in basic salaries	Not available
2. Travel allowances[b] (T A)	Rs 2.5/mile since 1984/85	Rs 101/day with 3-4 porters — 1985/86 to 1986/87 Rs 121.50/day with 3-4 porters — 1987/88 to date
3. Per Diem allowance[b]	Rs 35.00/day 1985/86 to present	Rs 366/day in Kathmandu: Rs 294/day elsewhere — 1986/87 to 1987/88 Rs 410/day in Kathmandu; Rs 330/day elsewhere 1987/88 to date
4. Provident Fund	10% Govt addition in 10% of personal deposit in base salary	10% addition in 10% of personal deposit in base salary — 1968/69 to 1977/78 1978/79 onwards only applied to HMG deputed staff
5. Gratuity	0.5, 1.0 and 1.5 months of salary for service years exceeding 5, 10 and 20 years respectively	Following applied from 1976/77 to 1987/88: 1/2 month pay for each full year of service — >5 to <10 years of service + 10% of total earning during the service period; 1 month pay for each year of service based on redundancy (resignation) — applied only when >1 year service; 1 month pay for each full year of service + 10% of total earnings
6. Pensions	Rs 580.0/month from 1984/85 to date	Not available
7. Dasai Bonus	1 month of basic salary	One month salary equivalent, payable only from July 1988
8. Health Insurance/ Treatment Allowance	12 months of salary for the whole of service period[c]	For employee, all expenses are reimbursed except for dental and optical. From July, 1988, same allowances are extended to wife and children but only when they are based on station. Doctor's certificate necessary. If hospitalised, dependents get only 50% expenses reimbursed.

Table 10 continued

Particulars	HMG	LAC/PAC
9. Children's Allowance	Only on death of the employee in service	Not available
10. Transportation	Plane fare, taxi fare and other transport against receipts	Sectional Head gets plane fare, taxi fare and other transport against receipts
11. Leave	Home leave — 30 days/ year Sick leave — 15 days/year Special leave without pay — 30 days Extraordinary leave without pay — 3 years Study leave (if in HMG's interest) — 5 years	Annual leave — 30 days Sick leave — 12 days/year Leave without pay — 3 months
12. Paid Holidays	—	16 days
13. Working Days/ week	6 days	5.5 days

a Remote areas include most of the difficult districts in the hills as specified in Fiscal Rules and Acts.

b 25-75% additions are provided in HMG T A and D A for remote and city areas.

c Difficult to obtain.

for specific import requirements. Second, LAC and PAC may increasingly be bound by agreements made centrally for germplasm exchange and testing with international organisations. LAC and PAC testing of, for instance, rice germplasm arriving through programmes agreed with IRRI, have generally failed to identify varieties that out-perform those available locally, and a recent external adviser's report (Willey, pers. comm.) recommends that this programme be reduced. It is conceivable that in future broader, national level, agreements made with international organisations through a central organisation such as NARCC, will be less sensitive to the types of interaction likely to be of benefit under specific agro-ecological conditions.

Perhaps the most pervasive potentially negative influence on LAC and PAC performance may arise through integration with HMG(N) administrative and staffing procedures. As noted above, LAC and PAC have hitherto enjoyed independence in precisely those day-to-day procedures which reviews have shown to function inefficiently under HMG(N), such as ordering procedures for consumables, and maintenance of buildings, equipment and vehicles.

Staff emoluments will also decline in real terms as HMG(N) conditions begin to apply. Data in Table 9 indicate a 3-fold difference in salary between HMG(N) and LAC/PAC staff of similar grade. Negotiations between HMG(N) and the British Government over the future of LAC and PAC are sensitive to the disincentive effect that a deterioration in employment conditions would bring. Current proposals envisage the phasing out of salary payments made hitherto directly from the UK under the Joint Services Scheme for Civilian Employment in Nepal (JSSCEN), the introduction of one-year (renewable) contracts for Nepali staff with basic salaries payable at equivalent HMG(N) levels. UK funding will enhance these basic payments to compensate for loss of allowances and pension rights which would have been accumulated had the employee been a member of HMG(N) service. A lump sum payment will be made as compensation for the discontinuation of JCSCEN, and an annual gratuity will be paid. The combined effect of the gratuity and other additional benefits is to raise total emoluments to a level comparable with that under JSSCEN. From the viewpoint of the employee, the main disadvantages of the new arrangements are the short-term nature of employment contracts, and the fact that basic salary will increase only in line with HMG(N) provisions.

The gradual withdrawal of UK support will, under the new arrangements, require merely the reduction of such additional benefits as the gratuity, and not adjustment to basic conditions of employment, and so is clearly facilitated. Current estimate suggest, however, that a withdrawal will be phased over a 10-year period in an effort to minimise disruption.

In parallel with the reduction in financial support, current proposals also envisage reduced technical assistance. For instance, of the six expatriate advisers currently at PAC, three (in social anthropology, forestry/pasture and livestock) will not be replaced after 1990, the socio-economist and director will continue to 1993, and farming systems adviser to 1995. A similar pattern will apply at LAC, a Nepali director being appointed in 1993 and assisted by only an expatriate research adviser from then to 1995.

At the same time, however, steps are being taken under the 1989-95 funding proposals for LAC and PAC to strengthen administration and develop new types of technical support in preparation for integration with HMG(N). An Assistant Director (Administration) is to be appointed from 1990 at PAC to allow the director to become more closely involved in overall strategy and in research management. On the technical side, as the numbers of long-term advisers decline, proposals are being made at PAC for links with institutions in the UK and elsewhere capable of supporting research programmes through the supply of bibliographic material, training and short advisory visits. The Memorandum of Understanding between HMG(N) and the British Government which provides the framework for 1989-95 funding for LAC and PAC also envisages the possibility of additional technical support at national level in research management and strategy, in the form of an expatriate adviser to be assigned to NARSC. Some discussion has also taken place of the possibility that technical and financial support for LAC and PAC might be requested from other donors in specified areas over the next decade. Overall, whilst those negotiating the future of LAC and PAC have been sensitive to the reductions in research productivity that could result from their loss of special status, much will depend on whether the ethos of LAC and PAC can survive changing conditions and whether other components of the Nepal research service will be capable of capitalising on the changes introduced at LAC and PAC. Sensitive monitoring of the transition by research managers and funding agencies will be needed if productivity is not to decline.

A Case study from Bolivia: collaboration between the Centro de Investigación en Agricultura Tropical (CIAT) and British Tropical Agriculture Mission (BTAM)

Philosophy and organisation

The chronology of principal events influencing decisions on agricultural research and extension in the Santa Cruz Department is given in Chapter 4. Briefly, following an initial period of stable funding for research and extension with US support (1948/61) when the agricultural potential of the Department was first being opened up, activity declined as the Ministry of Agriculture became the principal funding source in 1961-70. The discovery of petroleum in the Department in the late 1960s, the Central Government decision to establish decentralised Development Corporations (DCs) for the country's nine administrative Departments, and the agreement that these should be financed partly by taxes on resources exploited within the Departments (most importantly for Santa Cruz, on oil and gas) paved the way for Department-level initiatives on research and extension.

CIAT was established by Supreme Decree in February 1975, although financial resources were not allocated until the following year's budget. It was given the responsibility for research and extension for all farmers and all crops (except sugar cane, for which there is a separate research institute) within the Department, though in practice, its resources for extension during this early period were far fewer than for research, allowing the establishment of only three extension offices, against the (approximately) 20 remaining under the Ministry, and the (eventually) 15 'rural residents ' financed directly by the Departmental Development Corporation(DC).[4]

CIAT is headed by an Executive Director responsible to a Board of Directors, comprising representatives of the Ministry, the DC, the Departmental University, and the Chambers of Trade and Industry and of Agriculture. CIAT's current budget is approximately US $1m/year, 81% of which is provided by the DC, 10% by farmers' organisation, 2% by the Ministry and the remainder from CIAT's own resources (eg. sales of seed, soils analyses). The Board has — and exploits — the potential to influence the direction of CIAT research, though changes to date have been neither so frequent nor so major as to disrupt the flow of work. Levels of influence within the Board are roughly proportionate to the contributions by individual organisations to the CIAT budget. The Ministry's influence is therefore very limited, that of the DC paramount, and that of the Chamber of Agriculture (CAO) stronger than its level of funding might suggest, given that producers' organisations affiliated to it represent particularly the large and medium-scale farmers in the Department.

CIAT has technical cooperation agreements with a number of donor agencies. During the period 1985-89, these included: with the Japanese International Cooperation Agency (JICA) in rice, fruit and agricultural mechanisation; with FAO in fertiliser; with the Netherlands for research into soya inoculants; with USAID PL480 for seeds research and production, and with the Interamerican Development Bank and Interamerican Institute for Cooperation in

4. In Chapter 4 details are given of how, in 1988, overall responsibility for establishing and managing a system of extension was given to CIAT by the DC, with increased funding.

Agriculture for research and training in oilseeds, cereals and livestock.

The largest and most enduring collaborative programme has been with the British Tropical Agriculture Mission (BTAM). This was initially established as a team of up to 10 scientists with a national remit and based in La Paz from 1963 to 1972. The difficulties of coping with the wide diversity of agroecological conditions in Bolivia led to agreement with other donors in the early 1970s that the British effort would be based in Santa Cruz and concentrate on technologies for the lowlands, the Altiplano and Mesothermic valleys becoming areas of specialisation for USAID and FAO technical cooperation respectively.

BTAM was re-established in Santa Cruz in 1976 and, under the new name of British Sustainable Agriculture Programme, has recently been funded by ODA for a fifth period, from 1990-1995. The annual budget of BTAM is approximately £700,000 sterling. It has varied in size over the period 1976-1989 from four to eight professionals, all of whom have served on long-term contracts, the majority having stayed on for considerably longer than the two year minimum. BTAM's overall objectives have been first, to give technical support to CIAT in agreed programmes of research and in the overall management of research and (especially since 1988) extension and, second, to provide training to CIAT staff. This has taken the form of both in-service training through close counterpart 'twinning' within individual programmes, and courses (usually at MSc level, but increasingly also short courses) at institutes of higher education in the UK and in other Latin American countries. The impact of training by BTAM is assessed in Annex 6. Particularly worthy of note are the low turnover of staff in CIAT, and the fact that the great majority of those who do leave CIAT take up employment in other aspects of Santa Cruz agriculture.

Much of BTAM's early work was in helping CIAT to establish research programmes in the main commodities (maize, rice, livestock, pastures) and disciplinary (soils, weed and pest management, agricultural economics) programmes. Responsibility for the main commodity and disciplinary programme now rests largely with CIAT, some of them being pursued in collaboration with technical missions other than BTAM (eg. with JICA in rice) In the early 1980s, a review of BTAM urged closer concentration on technology development for the stabilisation of slash and burn cultivation by smallholders (principally internal migrants from the

Andes) and, simultaneously, BTAM began to introduce, and CIAT take up, more of a systems approach to research. A further review in 1989 recommended continued funding for BTAM until at least 1995 on the strength of its emerging contribution to environmental awareness in, first, the development of sustainable and profitable farming systems to replace shifting cultivation and, second, its increasing capacity to develop practical and low cost technologies to limit the damage done to the environment by mechanised farming.

Research methods

Through a process of evolution, CIAT/BTAM have developed or adapted methodologies to suit local conditions in three broad areas: a farming systems approach, development of off-station research facilities and adaptive research and dissemination in collaboration with intermediate organisations.

i) Farming systems approach: In their early periods BTAM and CIAT made substantial progress in addressing research problems relating to products of clear economic importance (pastures, cattle, rice, maize) and in filling obvious gaps in knowledge on a discipline basis (soil survey; pest, disease and weed control). Major adaptations of research methodology to suit local conditions came in the early 1980s on the initiative of the agricultural economics programme, when a farming systems approach to research was introduced, drawing on the 'sondeo' methodology of Hildebrand (1981).

This had been preceded by an extensive formal survey of farming systems in the Northern Colonisation Zone (NCZ), ie. that occupied by some 20,000 small-scale farmers recently settled in Santa Cruz Department from the Altiplano. This survey (Maxwell & Pozo, 1981), although in retrospect costly and time-consuming, generated research hypotheses for the NCZ, prompted a series of smaller studies to fill gaps in knowledge (Stutley, 1982; Maxwell et al., 1982) and stimulated an awareness that the components of small farm systems (annual crops, livestock, trees) were highly interdependent. It became clear that the search for 'avenues of escape' (Maxwell, 1979) from the factors causing slash and burn farmers to move on into virgin forest (weed infestation and poor regrowth of forest cover giving a poor burn) could only be pursued within a systems perspective.

Subsequent work led to the definition of FSR as it was to be applied by BTAM/CIAT, and the preparation of guidelines for its implementation. A number of rapid interdisciplinary field surveys were then conducted with the twin objectives of identifying research priorities for given research domains, and familiarising research staff with the practices and potential of the approach (Lawrence-Jones & Pozo, 1983; Lawrence-Jones et al., 1983; and Froment et al., 1984).

The successful implementation of farming systems approaches at field level encouraged efforts in CIAT to make its internal structure more compatible with the systems approach. Up to 1986, CIAT was structured along conventional commodity and support programme lines. However, as efforts originating from the agricultural economics unit aimed at the design and conduct of research within a systems context gained wide acceptance, it gradually became clear that a commodity-based structure was inadequate to capture important systems interactions. One of the first tasks of the CIAT Sub-Director newly appointed in 1986, therefore, was to propose structural alternations compatible with a farming systems approach. After lengthy internal discussion, a structure was introduced in which two broad agro-ecological zones were defined, multidisciplinary teams being assigned to each (originally proposed as four, but reduced for budgetary reasons). CIAT staff then reviewed existing knowledge of the farming systems for each zone, followed by the allocation of research staff to each according to discipline and experience, and by further discussion to identify the principal subjects on which research was required. At the same time a consultative process was initiated with Producers' Organisations, both by meetings and correspondence, in order to obtain their views on the priorities for research and, where possible, to incorporate them in the 1987-88 programme.

In spite of its negative implications for CIAT as a whole, the reduction by almost some 25% of CORDECRUZ's funding for CIAT in early 1987 (ie. from US $1.05m to US $0.76m) undoubtedly accelerated the move towards a problem-oriented and systems-focused research strategy, those programmes incompatible with this approach being the first to be threatened with cuts. In the event, eight research staff were made redundant in July 1987, one each leaving the following sections: phytopathology, wheat, pastures, livestock, oilseeds, rural sociology, soils and the San Pedro Local Experimental Unit.

ii) Off-station research facilities: The main effort has been to develop local experimental units, but on-farm trials are playing an increasingly important role.

The 1984 Review of BTAM, whilst recognising the substantial progress made since 1980 in introducing a farming systems approach, made specific recommendations for strengthening its implementation. One of these was for the establishment of LEUs, Local Experimental Units) as 'intermediate' sites between on-farm and main experiment station (Saavedra) levels where trials could be conducted under conditions similar to those on-farm, but a stronger degree of control would be feasible, thereby reducing the proportion of trials subject to uninformative failure, which had been particularly high in on-farm trials in the early 1980s. Taking up this proposal, Lawrence-Jones proposed (July, 1985) that four LEUs be set up in the Northern Colonisation Zone (NCZ). The philosophy and proposed organisation of LEUs is outlined in a joint CIAT/BTAM publication (CIAT/BTAM, 1985). The LEUs are rapidly becoming multipurpose, in addition to their adaptive trials functions serving also as retailers of generic material, focal points for visits by research and extension staff from NGOs and farmers' organisations and as centres for training courses and farmer discussion groups.

iii) Adaptive research and dissemination in collaboration with intermediate organisations: Non-governmental organisations have been taking initiatives in research and extension over a number of years, frequently in informal collaboration with CIAT/BTAM researchers. As mutual confidence has grown, contacts have been formalised into a system for adaptive research, extension and feedback incorporating NGOs as one of several types of 'intermediate organisations' between farmer and researcher, with finance from the Departmental Development Corporation (Thiele et al., 1988). Details are given in Chapter 4.

Progress in technology generation and the contribution of exotic genetic material

The long distances separating areas of cultivation in Santa Cruz, together with the wide range of farm types, make it more difficult to quantify the spread of improved genetic material and the importance of the CGIAR centres as sources of material there than in the Nepal case studies. As in Nepal, the proportion of CIMMYT genetic material in improved wheats is high, and moderately so in maize. CIAT (Colombia) has been a source of material for one of the more popular varieties of maize (Swan Saavedra) but so has farmers' own material. CIAT (Colombia) has had an important input into the most popular variety of rice (CICA-8) but material from IRRI is unimportant. All varieties of soya originate in the USA and Brazil and were tested by CIAT on introduction. Data are unavailable for pastures, but anecdotal evidence suggests that the influence of genetic material from CIAT (Colombia) is substantial. Overall, the influence of material from the CGIAR centres on crop improvement has been much higher than in Nepal. Firm estimates are impossible, but it appears that very approximately 50% of genetic material used in crop improvement originates in the CGIAR centres.

Analysis of two further programmes directed particularly towards small and medium-scale farmers — agroforestry and animal improvement — reveals, however, that genetic material is derived principally from locally-available but little utilised resources.

Financial and personnel management

Annex 6 notes the levels to which CIAT staff have been trained through the BTAM programme, and the rate of turnover of Bolivian professional staff. It concludes that turnover has been low — almost 50% of those employed by CIAT at any time since 1976 are still in post, the proportion reaching 60% of those trained to MSc level under the UK programme. This contrasts with an average annual level of resignation of some 20% of staff employed by the national research service, IBTA. Of those leaving CIAT, all but two remain employed in agriculture within the Santa Cruz Department, so that their professional training is still being put to use in the area intended.

Table 11: Salary comparisons for middle-level professional, CIAT and IBTA, Bolivia (US $-equivalent)

Year	*IBTA US$/month*	*CIAT US$ month*
1981	367	318
1982	56	83
1983	47	30
1984	12	103
1985	19	80
1986	46	n.a.
1987	50	676
1988	60	676

Note: The year-to-year variation is explained largely by varying rates of depreciation of the peso salary in relation to the dollar. See also Table 4.

Part of the reason for greater stability among staff at CIAT than at IBTA appears to be revealed by a comparison of salaries (Table 11), particularly in the last 2 years. However, both suffered severe erosion during the 1983-85 hyperinflation, and although firm data are unavailable, anecdotal evidence suggests that there was little difference in salary levels between the two organisations in the mid to late 1970s. It is therefore unlikely that the entire difference in turnover rates between the two organisations is explained by salary differences. Although it cannot be proven, it seems likely that at least part of the difference is explained by better working facilities, prospects for training and higher levels of job satisfaction at CIAT than at IBTA, in brief, the more powerful work ethos often accompanying decentralisation, smallness of institutional size and dynamism.

4
A Comparison of Approaches to Technology Dissemination and Feedback

In the last two decades a debate has emerged on the form and functions of agricultural extension for small farmers in ldcs. Some donor agencies see tighter control of the extension system and improved technical capacity as major requirements and have promoted systems having strong administrative hierarchies, of which the Training and Visit system is typical. On the other hand, many practitioners, concerned with the large numbers of rural poor seeking livelihoods in difficult agricultural conditions, recognise a need for participatory approaches to technology development and dissemination if technology suited to the complex and diverse agro-ecological conditions in these areas is to be identified (Chambers et al. (eds), 1989; Farrington & Martin, 1988). The respective merits of the two broad approaches have been summarised by Sagar and Farrington (1988). Whilst T & V scores highly as a staff management system, in improving technical backup and providing regular contact between researchers and extensionists, it is costly to operate and, in practice, has rarely succeeded in generating adequate feedback from farmer to researcher. Particular disadvantages of T & V for the difficult conditions forming the focus of LAC/PAC and BTAM work include its tendency to place financial, administrative and technical control in a single, central location often remote from such difficult areas. Another is its insistence that extensionists should be divested of functions ancillary to the principal one of communication, such as the supply of farm inputs.

The approaches to extension at LAC/PAC and BTAM, although widely divergent, depart from the premises that:

- Links between centrally-located institutions or services and difficult areas are adversely affected by low staffing densities and poor communications. Knowledge of conditions in these areas therefore tends to be poor, and the basis for identification of suitable technology weak. Approaches to agricultural extension must therefore be decentralised in management and, in their methods, institutions and personnel, tailored to local conditions.

- Poor communications mean that input supply is unreliable and expensive, and extension messages not backed up by the requisite inputs lack credibility. It is therefore impossible to separate extension from input supply functions. It is also crucial that reliance on purchased inputs be kept to a minimum.
- Participatory approaches are essential but expensive in terms of professional staff time. Methods of spreading or reducing these costs without loss of effectiveness must therefore be sought and incorporated into the extension system.

Approaches to extension now at Pakhribas

Prior to 1988, PAC's Extension Section worked largely with the Agronomy Section on crop-related technologies; thereafter it became responsible for the extension of all agricultural technologies in the Local and Northern Target Areas. These two areas — the former comprising 4163 farming families on 8068 ha of land in Dhankuta District, the latter, 4900 farming families on 10,000 ha in Terathum and Taplejung — have been the geographical focus of PAC extension efforts, although its remit as a research and resource centre has been wider, namely to serve the four districts of the Koshi Hills Development Programme and (from 1989) a further seven hill Districts in the Eastern Region.

The approach to agricultural extension that has evolved at PAC is characterised by:

- Integration of information and input supply. Given that credit is extremely limited in supply and, at best, accessible only to wealthier farmers, PAC's general approach has been to develop technologies which do not require external inputs. Where inputs are necessary (eg. fertiliser for wheat), PAC has occasionally purchased the necessary stocks and provided them when supplies in the local market have been uncertain (Thapa et al., 1988). Farmers' requirements of improved seed were initially met from PAC supplies, but local farmers have successfully been encouraged to take on this role. For instance, all wheat seed is now produced locally, seed production having been stimulated initially by the payment of porterage from the farm to PAC and, more recently by a quality-based premium of up to 30%. An open pollinated crop such as maize requires a different strategy. Here, villages were identified in which all the maize grown is of a single improved variety, and farmers contracted to produce pure seed.

- Training of farmers in techniques conducive to self-sufficiency in input supply. These, again, have focused on seed production techniques, and have included training in seed selection, treatment, storage and germination testing. As these skills are acquired, the responsibility for seed supply is gradually shifted from PAC to seed production groups.
- Community participation in extension. This finds expression in many forms, including the requirement for senior extension staff to spend a minimum of 80 days in the field per year, the maintenance of a small core of nationally recruited senior staff supported by junior staff recruited from villages within the target areas and the long hours of informal discussion and debate with farmers themselves. These processes have enhanced the accuracy of initial identification of technologies likely to be acceptable, and have contributed to an iterative process of improving technologies through further research and testing.
- Targeting of extension messages. In the last two years, disadvantaged ethnic groups, other poor farmers and women have been the focus of particular extension effort. An Extension Section trek focusing on women's activities not only provided contact with this hitherto largely neglected group, but also brought to the attention of PAC researchers the group's needs for particular kinds of technology and extension messages in eg. vegetable growing.
- Avoidance of technology 'packages' and rigid messages. 'The extension approach has been to demonstrate a range of alternative technologies to farmers and allow them to select the techniques that are appropriate for their particular circumstances' (Thapa et al., 1988). Although systems interactions are recognised, it has also proven useful to separate out the components of a technology, conducting separate on-farm testing and demonstration. This approach has the added advantage of allowing sequential introduction of innovations in step with farmers' capacity to incorporate them.
- An awareness of indigenous methods of transmitting new technology and the search for ways in which PAC assistance can support and stimulate them. The spread of 'pokhreli masino' rice from 1973-1986, is the best-documented case of farmer-to-farmer

dissemination (Green, 1987). Techniques such as 'focus visits' by groups of farmers from one locality to another are intended to stimulate informal processes of this kind.

Approaches to extension at Lumle Africulture Centre

LAC's extension philosophy derives from its early remit to provide training for resettled Gurkha soldiers — viz. it aims to serve farmers who demonstrate a wish to be assisted. These are, by definition, 'progressive', and whilst this may represent a cost-effective strategy for enhancing production, its distributional impact is impaired by the relative neglect of the majority of farmers outside this category, given the inadequacies of 'trickle down' (Crooks and Hughes, 1981). Much of LAC's on-farm experimentation and extension has concentrated on the 25 panchayats of the Extension Command Area in Parbat, Myagdi and Kaski Districts.

Seeley (1989) in particular notes that lower caste families tend to live some distance from village centres and so are less frequently visited by extension workers. She also notes the limited contact between male extension workers and women in farm households, and between members of different castes.

Approaches to extension at CIAT/BTAM

As outlined in the introductory chapter, agro-ecological and socio-economic conditions in eastern Bolivia differ markedly from those in Nepal: approximately 1000 large commercial enterprises located near Santa Cruz city coexist with semi-commercial ranching and small subsistence settlements in the remoter parts of the Santa Cruz Department, while substantial areas have been developed by foreign (Japanese, Mennonite) and national (from the high Andes) colonisation on the fringes of the Integrated Zone (Map 1, p.10; Table 3).

Most of the Department's 400 km of surfaced road is found in the Integrated Zone. Poor communications and scattered population (on average under 1 person per km^2 outside the city) have important implications for the organisation of extension services.

Since the beginning of large-scale settlement in Santa Cruz in the early 1950s, various models of extension have been tried. These include:

- 1948-1961: the Servicio Agricola Interamericano (SAI): USA funding supported the salary, transport and per diem costs of local researchers and extensionists. The SAI focused on the newly-established Integrated Zone, where substantial expansion in the main commercial crops (sugar, maize, rice) was achieved. The research-extension system was modelled on that of the USA in the early twentieth century, extensionists coordinating technical advice with input supply and credit.

- 1961-1970: direct budgetary support to the SAI ceased in 1961. Research and extension were incorporated into the Ministry of Agriculture. Reduced pay and poor conditions meant that almost all of the 57 research staff from Santa Cruz trained abroad in 1957-60 had left by 1963.
- 1970-1975: the emergence of a Departmental-level Development Corporation (DC) funded by resources from petroleum exploration led to the creation of several major agro-industrial projects. This body took over the funding of research and extension, the activities of which were geared to support local agro-industry.
- 1975-1983: extension was de-linked from agro-industry and placed under the newly established Departmental research station (CIAT). Intentions to spread extension services to small farmers were only partly realised, owing to shortages of funds. At the same time, the DC set up some 20 provincial offices geared to technology transfer. Although adequately funded, the technologies they were recommending lacked research backing and were introduced in a top-down fashion.
- 1983-1987: To address these difficulties, the DC established, with Spanish Technical Cooperation, a Centre for Rural Development. This eventually was abandoned as oil price crises reduced the DC's income.

In the light of these predominantly negative experiences, CIAT and the DC in 1987 recognised that the failure of publicly-funded extension was likely to be not a transitional but a chronic phenomenon in Santa Cruz, given its poor communications, low population densities and funding crises in the public sector. A consultant's report recommending the introduction of a modified version of Training and Visit had earlier been rejected for the same reasons, and in 1987 the DC agreed to formalise and fund technology transfer and feedback linkages which CIAT had for some years been developing with a diversity of agencies throughout the Department. Throughout this period CIAT has been assisted by the UK-funded British Tropical Agriculture Mission (BTAM). The informal linkages developed by CIAT/BTAM included:

- Producers' Organisations (POs): over ten in number, these represent each of the major commodities and are generally funded by a levy on sales. Their functions include input and credit supply, and technology transfer. In aggregate they employ 18 professional agriculturists and veterinarians.
- Non-governmental organisations (NGOs): both indigenous and foreign-based organisations have developed long-term rural development bases in even the most remote communities. Eight large NGOs employ 190 outreach staff, including 24 agriculture graduates and 23 technician level agriculturists over 24 locations, and have approximately 1500 families in the locations they serve participate directly in their programmes, a further 1200 having indirect contact. For the colonisation zone, the density of NGO activity is high — over 25% of all families have direct or indirect contact with their programmes, whilst for the more sparsely populated remainder of the Department, the proportion is lower, at 15% — 20%. NGO agriculturists undertake adaptive trials in collaboration with CIAT/BTAM.
- Area-based publicly-funded programmes: These include the Pladerve rural development project on the eastern border which has eight agriculture professionals working with 700 families.
- The continuing DC provincial offices, now 15 in number, employing 39 staff of various grades with technology transfer responsibilities, and having contact with over 5000 families.
- Private sector agencies, particularly agro-chemical and machinery companies. CIAT/BTAM conducts product registration trials for these and conducts trials on types of implement to suit varying soil conditions.

Given low levels of funding and the poor infrastructure of Santa Cruz, CIAT/BTAM recognises its inability to reach farmers directly and has now identified these diverse 'intermediate users' of its technologies as its main clientele. It has developed a multiplicity of extension approaches to match this diversity:

- For the small farm areas, three Local Experimental Units (LEUs) have been established to serve for adaptive trials, demonstration and sale of planting material. Two more are planned.
- Two NGOs participate regularly in CIAT/BTAM's programme of adaptive trials. Others try out CIAT/BTAM material informally. NGOs are invited to contribute to the annual planning meeting of CIAT's research programme.

Recent recognition by the DC of the validity of this approach opens the way for programme funding, both from the DC itself and from long-term UK Technical Cooperation, to finance publication and distribution of materials from CIAT to suit the requirements of its various clients. Initially, the focus is on:

- for small farmers, information on 'escape routes' from the slash and burn crisis, centring on the combinations of pasture, tree crops, windbreaks, livestock and (small amounts of) annual cropping developed through CIAT/BTAM research and now operational on pilot farms.
- minimising the possible ecological damage from large-scale mechanised farming. Soil management and windbreaks are of special importance here, and explicit recognition of CIAT's philosophy of working through 'intermediate users' has been given by a World Bank/FAO Investment Centre mechanised farming project which will establish and fund two further LEUs to collaborate with CIAT/BTAM in technology testing and dissemination.

In summary, the CIAT/BTAM philosophy is based on five arguments:

- That an orthodox, monolithic publicly-funded extension system is unworkable given the conditions in Santa Cruz.
- That local-level intermediate users have to be identified, having the funds and technical personnel to carry recommendations out to the farmers (in some cases after local modification).
- That a multiplicity of extension methods and materials has to be developed to suit these diverse clients.
- That these clients' local knowledge and their responsibility to local groups stimulates feedback and 'demand pull' on the research agenda.
- That public (sometimes foreign assisted) area-based projects can incorporate units (such as LEUs) for local trials and extension compatible with this model and can fund both these and their links with CIAT.

Finally, in addition to its technical material, CIAT will put out material describing its approach to adaptive research and extension. This is to encourage funding agencies (whether NGOs or aid donors) to design their extension efforts in a way compatible with this approach.

5
Conclusions

This book is not alone in arguing that technologies can be identified for sustainable agricultural development in fragile areas and that farmers play a key role in modifying, disseminating and adopting these technologies. Powerful arguments along the same lines have been advanced in two other recent books (Chambers et al. (eds), 1989; Conway and Barbier, 1990) and have long been central to other analysts (Richards, 1986; Harwood, 1979). What is specific to this book is the search for viable institutional forms that allow the flexible, innovative and participatory approaches essential to technology development in fragile areas to be put into practice on a large scale.

The prospects for success in institutionalising these approaches depend partly on the types and volume of resources available for technology development and dissemination in CDR areas and, on the ingenuity of researchers and research managers in matching these to research methods and to farmers' technology requirements.

Conclusions and recommendations are presented below in each of these areas.

Availability of publicly-funded resources for CDR areas

In both Nepal and Bolivia, government funding of the national research system is low, and the effects of this are particularly acute in CDR areas. The specific results of low funding include high staff turnover in Bolivia, and a high level of 'moonlighting' in Nepal. In both cases, motivation is low, and resources for fieldwork are severely limited.

Both countries have attracted substantial levels of donor funding which have reduced these problems to some degree, but donor programmes have tended to focus on specific issues or geographical areas. Research into potential alternatives to coca provide an example in Bolivia, as does donor assistance in the overall institutional strengthening of the national research organisation in Nepal.

In both countries, specific collaborative efforts between local agencies and donors have concentrated on technology development for CDR areas. This book has analysed the more illuminating of these. Such efforts — particularly in Bolivia — have recognised the importance of other agencies in making resources available for technology development in CDR areas, and have sought over a long period to collaborate with these agencies, ranging from NGOs to private commercial firms, allowing each to work in its field of greatest comparative advantage and so promoting an efficient allocation of research resources for the area as a whole.

Both shortages of data and conceptual difficulties make it impossible to construct and verify an economic model to optimise allocation of government resources among CDR and other areas. From circumstantial evidence, a case can be made for higher allocations to CDR areas, but the limitations of budgetary resources in developing countries are a reality and it must be accepted that any increases in allocations are likely to be small. At the same time, however, evidence from one of the case studies suggests that more efficient use can be made of existing government funds by resisting the temptation to spread them too thinly. The management costs of operating a research unit under CDR conditions are high, so that units must be large enough to justify this cost, and to reduce the risk of inefficient or non-operation. The availability of donor resources for CDR areas is being influenced by some of this circumstantial evidence, including the increasingly accepted view that farmers are responsive to appropriate innovations, that more appropriate methods of estimating the changes in productivity in CDR areas than via output per hectare can and should be used, and that — as illustrated in this study — institutional forms can be developed which produce technology appropriate to farmers' requirements.

Whilst increased allocations of government resources to CDR areas are important, not least if these institutional innovations are to be sustained and spread, much of the argument of this book has been to encourage recognition that other resources for research and dissemination exist — not least those of the farmers themselves in CDR areas and should be fully utilised. Two strikingly different illustrations of the application of this principle are the promotion of farmer-based seed production and distribution in Nepal, and the incorporation of a range of 'intermediate users' of technology into the processes of experimentation, dissemination and feedback in Bolivia.

Institutional innovations for CDR areas — evidence from the case studies

The case studies are drawn from widely contrasting environments — the mid-Hills of Nepal on the one hand, and the eastern lowlands of Bolivia on the other. There are major differences in the overall strengths and weaknesses of the national research and extension services in the two countries, and in the form and status of donor assistance specifically intended to support research institutions for CDR areas.

Despite these differences, certain commonalities of approach emerge:

- Stable funding of the technical cooperation effort over a long period, with emphasis on long-term inputs from expatriate staff. Annex 7 illustrates that the 15 expatriates employed on BTAM since 1976 have served for an average of 4.5 years, and six remain in post.
- Training for local staff, whilst working in mixed local-expatriate teams, and in formal courses; learning by expatriates from local staff, expatriates drawing selectively on experience accumulated from elsewhere.
- An open attitude on the part of both local and expatriate staff, including awareness of the need to learn from farmers the complexities of managing local production system.
- A willingness to spend long periods in the field, working in teams of mixed disciplines and using a mixture of research methods, ranging through rapid appraisals, in-depth studies and formal trials.
- A willingness to break out of orthodox patterns of collaboration with other organisations where these have been shown to be of limited relevance. In genetic resources research, for instance, germplasm exchange and testing linkages with the international research centres have recently been much reduced in scope in Nepal and are exploited only for selected commodities in Bolivia. The corollary of this is increased focus on local genetic material (rice in Nepal, criollo cattle in Bolivia, trees in both countries) with the introduction of exotic germplasm only where proven to result in a product more acceptable to the farmer. Local resources

of a different kind — farmers' willingness to experiment; farmer-to-farmer networks; the skills and financial resources of intermediate organisations — have been used in both cases to supplement the resources of research centres.

Two preconditions for success have existed in both cases:

- A higher level of resources for staff remuneration, materials and equipment than is the norm for CDR areas.
- A high degree of decentralisation in day-to-day management and in strategic and policy decisions.

We return to both of these below.

Policy implications

For developing country governments

Farmers in CDR areas *are* responsive to new technology providing that it is appropriate to their requirements, is low-cost, and does not require major investment in new skills. Guidelines are now available from long-term practical experience on the types of institutional arrangement permitting cost-effective implementation of research methods intended to produce such technology. There is therefore a case for higher funding of technology development in CDR areas than in the past. If governments themselves are unable to provide increased funding, there is no reason why the resources which are available from government should not be supplemented by others from donors, NGOs and local organisations. How to combine such diverse resources effectively has been a major focus of the case studies reported here. In all events, governments should resist the temptation to spread their resources too thinly: research in and for CDR areas requires a high calibre of staff, strong motivation and adequate resources for field work especially. These conditions will not be fulfilled unless resources are concentrated to provide attractive living and working conditions, with levels or renumeration and allowances higher than have historically been provided from government alone. Finally, forms of decentralisation need to be found allowing CDR research institutes the necessary freedom of action in both day-to-day management and in decisions on research policy and strategy.

For donors

The sensitive use of donor funds over a long period, and long-term inputs from technical cooperation staff are essential for success in the paramount objective building a local institutional capacity to develop technology acceptable to CDR farmers. As the contrasting patterns of donor involvement in the case studies demonstrates, there need be no set format for donor involvement in institution building: in the Bolivia case, donor support was linked to an existing (decentralised) institute; in Nepal, new centres were established, almost entirely with donor funding, but with the objective of integrating these eventually into the government service.

The personal qualities for successful involvement of expatriate personnel are principally attitudinal and therefore difficult to teach, and include the views that farmers and local colleagues are 'equals', and that there is much to be learned about local farming practice.

For international research centres

Important lessons concern the limited relevance of much of their genetic material and of short term consultancies to the development of CDR areas. Indeed, far from pursuing application of the 'central source' model of technology development (a course of action implicit in the recommendations of Mellor (1988)), the international centres need comprehensively to redefine their role in CDR areas in line with the 'multiple source' model (Biggs, 1989). This implies that sources of innovation in material, methods and institutions exist outside the international and national research services, and are successfully introducing and disseminating technology in CDR areas both from their own resources and in collaboration with public sector agencies. Recognition of this reality will stimulate the international centres to seek and deploy their comparative advantage in relation to other agencies operating in CDR areas. Again, there are no blueprints, but the evidence suggests that their advantage will lie less in genetic material or short advisory consultancies, and more in the provision of appropriate training courses and in developing and legitimising innovative research methods. Certainly, as far as plant improvement is concerned, the evidence suggests limited relevance of genetic material from the CGIAR centres even among the field crops in which they have long specialised. Among those crops in which, from the case studies, they do appear to have some suitable genetic material (wheat, maize, rice) there appears to be a continuing need to train ldc plant breeders in how to draw selectively on the centres' resources in ways compatible with their breeding programmes, and to provide strong feedback to the international centres. Resources in CDR research are altogether too scarce to allow poorly-focused approaches to continue. For trees, livestock and the management of soil and water resources, which constitute a high proportion of the activities of research and extension for the CDR ares, there is no evidence from the case studies that ideas or materials emanating from the centres had any influence on the development of approaches or technologies.

Bibliography

APROSC (undated) 'Evaluation of the Integrated Cereals Project'. Draft, Kathmandu.

Balogun, P. 1989 'A review of published agricultural statistics in the Western Development Region of Nepal'. *Technical Papers* No 89/Lumle Agricultural Centre, Lumle.

Balogun, P. (in preparation) 'A review of methodological issues in sampe surveys of agriculture in Nepal'. Kathmandu: LAC.

Basnayat, T.B. and Yazman, J.A. 1988 'Problems and prospects of livestock research activities'. Draft, mimeo. Kathmandu: ARPP.

Biggs, S.D. 1980 'Informal R & D: farmer resistance to new technology is not always a sign of backwardness'. *Ceres*, 13 (4), pp 23-26.

Biggs, S.D. 1989 'A Multiple Source of Innovation Model of Agricultural Research and Technology Promotion'. *Agricultural Administration (Research and Extension) Network Paper* No. 6. London: ODI.

Biggs, S.D. and Farrington, J. 1990 *Agricultural Research and the Rural Poor: A review of social science analysis.* Ottawa: IDRC.

Birdsall, N. 1985 'Population growth and structural transformation of the labour force'. Paper for the Annual Meeting of the Population Association of America, Boston, 28-30 March.

Chambers, R.; Pacey, A. and Thrupp, L.A. (eds) 1989 *Farmer First: farmer innovation and agricultural research.* London: Intermediate Technology Publications.

Chambers, R. and Ghildyal, B.P. 1985 'Agricultural research for resource-poor farmers: the farmer-first-and-last model'. *Agricultural Administration and Extension*, 20, pp 1-30.

Chambers, R. and Jiggins, J. 1986 'Agricultural research for resource-poor farmers: a parsimonious paradigm'. *Discussion Paper* 220. IDS, University of Sussex.

CIAL 1988 'Comisión de Investigación en Agricultura Alternativa: Informe Seminario Latinamericano: Investigación en Agricultura Alternativa para el Desarrollo Campesino'. Santiago, Chile (mimeo).

CIAT/BTAM 1985 'Project proposal: a new structure for regional investigation, with proposals for the Northern Colonisation Zone'. *Working Document* Series.

Conway, G.R. and Barbier, E.B. 1990 *After the Green Revolution: sustainable agriculture for development*. London: Earthscan.

Crooks, G.R. and Hughes, D.T. 1981 'A Report on the Socio-Economic and Extension Sections at Lumle Agriculture Centre'. (mimeo). Kathmandu: LAC/PAC office.

de Boer, J. 'Technical assistance activities under ARDP during 1988'. Kathmandu: ARPP (mimeo).

Farrington, J. and Martin, A. 1988 *Farmer Participation in Agricultural Research. AAU Occasional Paper* No 9. London: ODI.

Farrington, J. and Biggs, S.D. 1990 'NGOs, Agricultural Technology and the Rural Poor'. *Food Policy*, 1990, Vol. 15, No. 6.

Freemen, W.H. 1987 'Strengthening agricultural research management'. ARPP Consultancy Report No 10. Kathmandu.

Froment, M.A.; Lawrence-Jones, W.; Solares, G. and Rodriguez, N. 1984 'Research Planning: the Farm Systems Project, Chané Pirai'. *Working Paper* No 43. Santa Cruz: CIAT/BTAM.

Galt, D. and Mathema, S.B. 1987 'The *Samuhik Bhraman* process in Nepal: a multidisciplinary group activity to approach farmers'. Paper presented to the Workshop on 'Farmers and agricultural research: complementary methods' IDS, University of Sussex, 26-31 July 1987.

Gatenby, R.M. and Shrestha, N.P. 1989 'Pakhribas pigs — an improved technology for poorer farmers'. pp. 29-32 in *Agricultural Administration (Research and Extension) Network Newsletter* 21. London: ODI.

Green, T. 1987 'Farmer-to-farmer seed exchange in the Eastern Hills of Nepal: the case of 'pokhreli masino' rice'. PAC Working Paper 10/87. Pakhribas Agriculture centre, c/o P O Box 106, Kathmandu: LAC/PAC Office.

Government of India 1985 *The Seventh Five Year Plan, 1985-90*. Government Printer, New Delhi.

Harwood, R.R. 1979 *Small farm development: understanding and improving farming systems in the humid tropics*. Boulder: Westview Press.

Hildebrand, P. 1981 'Combining disciplines in rapid appraisal: the sondeo approach'. *Agricultural Administration*, Vol. 8, pp. 423-432.

IBTA 1987 'Plan operativo 1987-88'. La Paz: MACA/IBTA.

ISNAR 1989 'Fortalecimiento del Sistema de investigación y transferencia de technología agropecuaria en Bolivia'. Report prepared for Government of Bolivia. The Hague: ISNAR.

KC,; Sharma, A.R.; Mathema, S.B. and Wagle, M.P. 1987 'Effectiveness and performance survey report of Kabre agriculture farm'. Report No 11. Kathmandu: SERED.

Kayastha, B.N.; Mathema, S.B. and Rood, P. 1989 'Nepal: organisation and management of on-farm research in the national agricultural research system'. *OFCOR Case Study* No 4. The Hague: ISNAR.

Khadka R. and Gibbon, D. 1988 'The contribution of Pakhribas Agricultural Centre to the development of farming systems in the eastern hills of Nepal'. *Technical Paper* No 111, Kathmandu: Pakhribas Agricultural Centre.

Lawrence-Jones, W. and Pozo, M. 1983 'A report on the implementation of a *sondeo* in the Mesomeric Valleys as a Method of Informal Reconnaissance'. *Working Paper* No 7. Santa Cruz: CIAT/BTAM.

Lawrence-Jones, W.; Froment, M.A.; Rodriguez, N.; Solares, G.; Martinez, L. and Encinas, C. 1983 'Prioridades para la investigación: un sondeo en el area de La Guardia-La Angostura'. [Research prioritisation: a *sondeo* in the La Guardia-La Angostura area]. *Working Paper* No 42. Santa Cruz: CIAT/BTAM.

Lewis, W.A. 1979 'The dual economy revisited'. *The Manchester School* 47 (3) September.

Maxwell, S. 1979 'Colonos marginalizados al Norte de Santa Cruz-avenidas de escape de la crisis de barbecho' [Marginalised colonists in the North of Santa Cruz-avenues of escape from the 'regrowth' crisis].

Maxwell, S. and Pozo, M. 1981 'Farm systems in the Colonisation Crescent of Santa Cruz, Bolivia: Results of a survey'. Vol I: text. *Working Paper* No 22. Santa Cruz: CIAT/BTAM.

Maxwell, S.; Stutley, C. and Bojanic, A. 1982 'Informe sobre un programa de estudio de casos en San Pedro' [Report on a case study program in San Pedro]. *Working Paper* No 27. Santa Cruz: CIAT/BTAM.

Mellor, J.W. 1986 'The new global context for agricultural research: implications for policy'. Address presented at International Centers Week, 3 November 1986, Washington and reprinted in *International Food Policy Research Institute Report 1986*, pp 7-14.

Mellor, J.W. 1988 'Agricultural development opportunities for the 1990s — the role of research'. Address to the International Centers Week of the CGIAR, Washington, DC, 4 November.

Mennonite Central Committee 1989 'Agriculture program research results, Bangladesh'. Annual Report No 16. Dhaka.

Oram, P. and Bindlish, V. 1981 *Resource allocations to national research systems: trends in the 1970s*. The Hague: ISNAR/IFPRI.

Pray, C.E. 1987 'Private sector agricultural research and technology transfer in LDCs: report on Phase II'. Department of Agricultural Economics, Rutgers University, March 2, 1987.

Richards, P. 1986 *Coping with hunger: hazard and experiment in an African rice farming system*. London: Allen and Unwin.

Röling, N. 1988 *Extension Science*. Cambridge: Cambridge University Press.

Sagar, D. and Farrington, J. 1988 'Participatory approaches to technology generation: from the development of methodology to wider-scale implementation'. *Agricultural Administration (Research and Extension) Network Paper* No. 2. London: ODI.

Seeley, J. 1989 'Women and extension services in Lumle Agriculture Centre Research and Extension Command Areas'. *Technical Paper* 16 (New Series). Lumle Agriculture Centre.

Sharland, R. 1989 'Indigenous knowledge and technical change in a subsistence society: lessons from the Moru of the Sudan'. *Agricultural Administration (Research & Extension) Network Paper* No 9 London: ODI.

Sharma, A.R.; Krishna, K.C.; Mathema, S.B. and Sharma Paudyal, B.K. 1988 'Effectiveness and performance survey report on Surkhet agriculture station'. Report No 13. Kathmandu: SERED

Sharma, R.P. and Anderson, J.F. 1985 'Nepal and the CGIAR Centres: A Study of their collaboration in agricultural research'. Washington DC: CGIAR.

Sthapit, B.R.; Pradhanang, P.M.; Subedi, K.D.; Sen, C.K.; Gurung, K.J. and Balogun, P. 1988 'Lumle report on the maize crop'. Report No 88/6. Kathmandu: LAC.

Stutley, C. 1982 'A study of the problems of destumping in the Chané-Pirai Colonisation area North of Santa Cruz'. *Working Paper* 34. Santa Cruz: CIAT/BTAM.

Thapa, H.B.; Green, T and Gibbon, D. 1988 'Agricultural Extension in the Hills of Nepal: Ten years of experience from Pakhribas Agricultural Centre'. *Agricultural Administration (Research & Extension) Network Paper* No 4. London: ODI.

Thiele, G.; Davies, P. and Farrington, J. 1988 'Strength in diversity: innovation in agricultural technology development in eastern Bolivia'. *Agricultural Administration (Research & Extension) Network Paper* No 1. London: ODI.

Velasco, R.; Thiele, G. and Davies, P. 1988 'Estudio de la transferencia de technologia agropecuaria en el Departamento de Santa Cruz', Vol I; Vol II (Annexos). [Study of agricultural technology dissemination in Santa Cruz Department]. *Working Document* No 68. Santa Cruz: CIAT/BTAM.

Wise, W.S. 1981 'The theory of agricultural research benefits?' *Journal of Agricultural Economics*, 22 (2), pp 147-157.

World Bank 1986 'Poverty and Hunger. Issues and options for food security in developing countries'. A World Bank Policy Study. Washington: World Bank.

Yadav, R.P. 1987 'Agricultural Research in Nepal: Resource Allocation, Structure and Incentives'. *Research Report* No 62. Washington DC: IFPRI.

Annex 1: High-yielding cereal varieties released in Nepal, 1959-85

Serial No.	Variety	Origin	Year of Release	Recommended for Hill	Tarai	Remarks[a]	
Rice							
1	CH-45	China	1960	•	•	Early,	1
2	N-136	India	1962		•		0
3	BR-34	India	1962		•		0
4	BR-8	India	1962		•		0
5	Taichung-176	Taiwan	1966	•			2
6	Chainung-242	Taiwan	1966	•			3
7	Kaoshung-24	Taiwan	1966	•			0
8	Chainan-2	Taiwan	1966	•			0
9	Tainan-1	Taiwan	1966	•	•		0
10	Taichung Native-1	Taiwan	1967		•		3
11	IR-8	IRRI	1968		•		0
12	IR-5	IRRI	1969		•		0
13	IR-20	IRRI	1970		•		0
14	IR-22	IRRI	1972		•		0
15	Khumal-1	Taiwan	1972	•			0
16	Parwanipur-1	Taiwan	1973		•		3
17	Jaya	India	1973		•		3
18	Masuli (Mahsuri)	Malaysia	1973		•		1
19	IR-24	IRRI	1975		•		0
20	Chandina (IR532)	IRRI	1977		•	Early,	3
21	Laxmi (IR2061)	IRRI	1979		•	Early,	2
22	Durga (IET2938)	India	1979		•		3
23	Sabitri (IR2071)	IRRI	1979		•		1
24	Janaki (BG-90-2)	Sri Lanka	1981		•		1
25	Bindeswari (IET1444)	India	1981		•	Early,	1
26	Kanchan (IR3941)	IRRI	1982	•			1
27	Himali (IR2298)	1982		•		Early,	1
28	Malika (Mala/J15)	Bangladesh	1982	•			2
29	Khumai-3 (CH 1039XIR580)	India	1983	Valleys		Early,	2
30	Khumal-2	Nepal	1987	Kathmandu Valley and Hills			1
31	Khumal-4	Nepal	1987	""			1
32	Palung-2	Nepal	1987	•			2
33	Chaite-2	IRRI	1987		•	Early,	2
34	Chaite-4	IRRI	1987		•	Early,	2
35	Barkhe-2	Bangladesh	1987		•		
36	Makwanpur-1	Sri Lanka	1987		•		2
37	Khajura-2	India	1987		•		2
38	Ghaiya-2	India	1987		•		

Serial No.	Variety	Origin	Year of Release	Recommended for Hill	Tarai	Remarks[a]
Maize						
1	Amarillo de Cuba	USA	1959	•	•	Early, 2
2	Cubano Flint	USA	1959	•	•	2
3	Khumal Yellow		1968	•		1
4	Rampur Yellow		1968		•	2
5	Kakani Yellow		1969	•		2
6	Hetauda Composite	Nepal	1972	•	•	3
7	Rampur Composite	Thailand	1975	•	•	1
8	Sarlahi White	Philippines	1975		•	3
9	Janaki Makai		1979		•	Winter, 3
10	Arun-2	UNCCXDHR	1981	•	•	1
11	Makalu-2	(Amarillo del Bajio)	1984	•		1
12	Manakamana-1	Nepal	1987	•		1
Wheat						
1	Lerma-52	CIMMYT	1960	•		2
2	Petic-62	CIMMYT	1967	•		0
3	Lerma Rojo-64	CIMMYT	1967	•		3
4	S-331	India	1969		•	0
5	S-227	India	1969		•	0
6	RR-21 (Sonalika)	India	1971	•	•	1
7	NL-30	India	1975		•	2
8	HD-1982	India	1975		•	0
9	UP-262	India	1979		•	1
10	Lumbini	India	1981		•	2
11	Siddhartha	India	1983		•	2
12	Triveni	India	1983		•	2
13	Vinayak	India	1983		•	2
14	Vaskar	CIMMYT	1983		•	2
15	Nepal 1297		1985		•	
16	Annapurna-1	CIMMYT	1988	•		Recent Rel
17	Annapura-2	CIMMYT	1988	•		Recent Rel
18	Nepa 251	India	1988		•	Recent Rel

[a]According to the FAO Seed Status Survey Report, the importance or popularity of varieties is categorised as follows:
1 = very important
2 = important
3 = less important
0 = grown occasionally or not at all.

Sources: Nepal, Department of Agriculture, *The Integration of Research and Extension in Farmers' Fields: The Terminal Report of the Integrated Cerals Project* (Kathmandu: Ministry of Agriculture, 1985), p. 49; and Food and Agriculture Organization of the United Nations, *Seed Status Survey Report* (Rome: FAO. 1981). *Seed Technology and Improvement Program (STIP)*, NARSC, Unpublished reports, 1989.

Annex 2: Physical Features of LAC and PAC Command Areas

LAC	PAC
Extension Command Area (1974-) Altitude: 917m - 3050 m Rainfall: range: 1358 mm- 5500 mm Mean:	*LTA* Altitude: Rainfall: range: Mean: *NTA* Altitudes Rainfall: range: Mean:
Extension Command Area (1974-) *Topography* Valley bottoms at 900-1000m Steeply sloping ridges at 2000m-3000m	*LTA and NTA (1975-)* As LAC, but valleys at 300-500m
Soils	eutric flavisols (300-800m) eutric cambisols (800-1700m) dystric cambisols (above 1700m)
Rainfall range 1350-5500mm/yr mean at LAC: mm/yr 70% of rainfall in June-September monsoon	range 900mm-2000mm/yr mean at PAC: mm/yr 70% of rainfall in June-September monsoon.

11 Districts (1989-)

Major changes will include:
shifting of the upper limits of altitude (to over 7000m)
and climatic (to alpine and arctic) conditions, with corresponding broadening of the range of farming conditions towards temperate crops and livestock.

Annex 3: Chronology of major developments, LAC and PAC

LAC	PAC
1968-74 Gurkha training in agriculture as part of resettlement programme	**1973-75** Gurkha training in agriculture as part of resettlement programme
1974 with ODA funding, mandate broadened to research, extension and training	**1975** with ODA funding, mandate broadened to research, extension and training
Work begun in 'Extension Command Area'(ECA) comprising 27 panchayats around LAC in Parbat, Myagdi and Kaski Districts.	Work begun in 'Local Target Area'of 7 panchayats adjacent to PAC and 'Northern Target Area' of 8 panchayats
ECA: area: 800 km^2 Population: 80,000 Cultivated area: 11,000 ha	*LTA, NTA*: area: Population: 55,000 Cultivated area: 7,000 ha
	1982 PAC broaden 'Research Command Area' by support for 4 Districts of KHARDEP (Dhankuta, Terathum, Bhojpur, Sankhuwasabha) comprising 6400 km^2, 109,000 ha of cultivated land and approximately 550,000 population

1985 Work of LAC and PAC formally incorporated into national agricultural research programme through NARSC

LAC	PAC
1986 LAC 'Research Command Area' broadened to include ECA plus 4 Districts of Hill Food Production Project (Lamjung, Gorka, Syangja,Tanahun).	**1987** PAC formally integrated into KMDP (successor to KHARDEP)

Annex 3: Chronology of major developments, LAC and PAC (continued)

	LAC	PAC
1989-95	(proposed) LAC 'Outreach Command Area' expanded to cover remaining panchayats of original 3 Districts, plus 4 HFPP Districts plus 4 further Districts	PAC to be separated from KHDP PAC 'Outreach Command Area' expanded to cover all 7 Districts of Mechi and Sagarmatha Zones, in addition to original 4 of Koshi Zone *Basic data for all 11 Districts* Area: 21,300 km^2 Population: 1.5 m Cultivated area: 300,000 ha

Annex 4: Farming Systems Data - LAC and PAC

	LAC (ECA + 4HFPP Districts) Annual Cropped area (ha)	*PAC (4 Koshi Zone Districts)* Annual Cropped area (ha)
1. *Main crops*		
Khet systems		
Rice		42549
Maize		7317
Wheat		252
Other cereals		13758
		63876
Bari systems		
Maize		63296
Millet		44582
Other cereals		1753
Potato		9641
Oilseeds		3108
Grain legumes		3985
Winter crops		510
Other		248
		127123
2. *Principal cropping patterns*		
Khet systems		
Rice - other cereal[1]	16770	18034
Rice - fallow	15311	16983
Rice - rice - fallow	2077	1884
Maize - rice - fallow	15209	7066
Maize - rice - wheat	2441	252
Bari systems		
Maize - millet[2]	58000	44521
Potato and maize	-	9132
Maize - oilseed[3]	-	7000
Maize - cereal	2500	1753
Maize - pulse	6000	-
Uplandrice - maize	1000	-

Sources: Sthapit et al. 88/28; Khadka and Gibbon, 1988

Notes:

1. Rice-wheat exclusively in Koshi hills, and predominantly elsewhere.
2. In Koshi hills, grown as two distinct sequential crops, but also in some areas as relay crops.
3. Includes interplanted maize and soya in Koshi Hills, and maize-mustard as distinct sequential crops.

Annex 5: Contribution of genetic material originating from CGIAR centres to crop production, research command area (4 districts), LAC

		Rice				Maize				Wheat		
		a	b	c = (a x b)		a	b	c = (a x b)		a	b	c = (a x b)
Principal Improved Varieties	Mahsuri	65	0	0	Ranpur Comp.	40	0	0	Sonalika	75	50	37.5
	Sabitri	15	50	7.5	Khumal Yellow	30	0	0	Triveni	10	50	5
	Janaki	2	10	0.2	Kakani Yellow	15	0	0	Lerma 52	5	0	0
	Kanchan	5	0	0	Arun-2	10	20	2	UP 262	5	0	0
	Himali	5	0	0	Others	5	0	0	Others	5	0	0
	CH-45	8	0	0								
	TOTAL			7.7				2				42.5

	Rice		Maize		Wheat	
Total area planted/yr in RCA		64,760	60,000		8,694	
Area planted to improved vars. (ha)	25,900	(40%)	33,000	(55%)	8,694	(100%)
Area-equivalent of genetic material from CGIAR centres (ha)	1994	(3.1%)	660	(1.1%)	3,695	(42.5%)

Annex 5: Contribution of genetic material originating from CGIAR centres to crop production, research command area (4 districts), LAC (continued)

	Millet				Barley				Potatoes			
		a	b	c = (a x b)		a	b	c = (a x b)		a	b	c = (a x b)
Principal Improved Varieties	Okhale-1	100	0	0	Bonus	100	0	0	Kufri Jyoti	75	0	0
									Lumle R8	20	0	5
									NPIT 0012	5	100	5
TOTAL				0				0				5

	Millet		Barley		Potatoes	
Total area planted/yr in RCA		184,200		1,200		1,500
Area planted to improved vars. (ha)	9,200	(5%)	540	(45%)	600	(40%)
Area-equivalent of genetic material from CGIAR centres (ha)	0	(0%)	0	(0%)	30	(2%)

NOTES: Column (a) gives the percentage of improved variety area under each variety specified
Column (b) gives the proportion of parentage of each variety derived from CGIAR centres' genetic material.
In column (c), by multiplying (a) and (b), an index is obtained of the improved variety area-wise significance of genetic material from the CGIAR centres. This is then translated into overall area equivalent for each crop in the final row.

Annex 6: The Training of Bolivian staff through BTAM

Until recently, lack of human resource development has been one of the factors limiting agricultural research in Santa Cruz. When members of the national-level BTAM worked there in the 1960s, there were few cruceños with even BSc-equivalent degrees, and counterpart staff were brought in from the Altiplano. The Faculty of Agricultural Sciences in the Santa Cruz Gabriel René Moreno was not set up until the early 1970s and even now produces BSc-equivalent graduates.

Since CIAT became operational in 1976, the current BTAM has played a major role in training in three broad ways:

1) on-the-job interactions between local and expatriate staff
2) BTAM support for BSc thesis work from the local community
3) higher degree and short course training abroad.

An internal review was carried out within CIAT/BTAM in 1988 to establish the level of turnover of Bolivian professional staff and to identify whether and how the training (and particularly the more formal type under (3)) was currently being utilised. Results of the review are summarised in Table A6.1. These indicate first, a slow overall turnover of staff in CIAT: almost half of those who worked as BTAM counterparts at any time since 1976 are still employed within CIAT, including the majority (8 from 13) of those who received MSc training. The average length of employment in CIAT overall has been 5 years, and for those with MSc, 7 years.

Apart from two former counterparts who could not be traced, all those not currently employed in CIAT remained in agriculture, overwhelmingly in Santa Cruz Department, with rather more in the private than in the public sector.

As far as a lower-level training is concerned, BTAM had in 1977-87 provided grants, supervision and material support for a total of 52 graduates from the local university to complete the experimental work on which their BSc theses were based. Of the 25 whose current employment is know, all had continued to work in agriculture, the majority in public sector agencies or in producers' organisations, again, exclusively within Santa Cruz Department.

Overall the conclusion drawn from this data is that counterpart staff turnover rates have been much lower in CIAT than more widely in Bolivian agricultural research institutes, that a higher proportion of those receiving postgraduate training have remained in CIAT, and that practically all former counterparts have continued to work in agriculture in Santa Cruz. Overall, the 'leakage' of BTAM-trained counterparts from CIAT has been acceptably low and almost all of those leaving have continued to apply their knowledge in Santa Cruz agriculture.

Table A6.1 Present (1987) employment of CIAT staff who have worked as BTAM counterparts at any time since 1976

	Those receiving UK funded MSc training	Total
Still employed in CIAT	8	17
Employed in other public sector agricultural agency	4	7
Employed in private sector agriculture	0	10
Working abroad in agriculture	1	1
Current employment unknown	0	2
Total	13	37

Notes: includes two who took courses but failed to obtain a qualification

Annex 7: Length of service of BTAM expatriate personnel as at March 1989

Name	Discipline	Duration	Person- Months	
C R Horrell	Pastures	1976 - 81	65	
J E Tollervey	Weed Control	1976 - 83	84	
J E Moore	Crop Protection	1976 - 80	44	
R O S Clarke	Crop Protection	1980 - 84	48	
R T Patterson	Pastures	1976 - 84	103	
J V Williams	Livestock	1976 -	164	continuing
S J Maxwell	Agricultural Economics	1978 - 82	48	
M Froment	Annual Cropping Systems	1984 - 86	23	
W Lawrence-Jones	Agricultural Economics	1982 - 84	48	
P Lee	Perennial Crops	1984 - 86	23	
R Barker	Soils	1984 -	65	continuing
P Davies	Agricultural Economics	1986 -	42	continuing
J Johnson	Agro-Forestry	1987 -	26	continuing
E Wyrley Birch	Perennial Crops	1987 -	32	continuing
G Thiele	Sociology	1987 -	35	continuing

Note: Two expatriates for whom data are not readily available are excluded.